# Dynamics of estuarine muds

## A manual for practical applications

Richard Whitehouse
Richard Soulsby
William Roberts
Helen Mitchener

Thomas Telford

Published by Thomas Telford Publishing, Thomas Telford Ltd,
1 Heron Quay, London E14 4JD.

URL: http://www.thomastelford.com

Distributors for Thomas Telford books are
*USA:* ASCE Press, 1801 Alexander Bell Drive, Reston, VA 20191-4400, USA
*Japan:* Maruzen Co. Ltd, Book Department, 3–10 Nihonbashi 2-chome, Chuo-ku, Tokyo 103
*Australia:* DA Books and Journals, 648 Whitehorse Road, Mitcham 3132, Victoria

First published 2000

A catalogue record for this book is available from the British Library

ISBN: 0 7277 2864 4



Typeset by Academic + Technical Typesetting, Bristol
Printed and bound in Great Britain by MPG Books, Bodmin, Cornwall

# Preface

This manual deals with the main processes of cohesive sediment behaviour, namely, erosion, transport, deposition and consolidation, and is based on the report of the same title produced within the HR Wallingford strategic research series (report SR 527). It updates, and replaces, the earlier editions of this manual published as HR Wallingford reports in 1988 and 1992 by incorporating the results of recent research undertaken by HR Wallingford, and includes the results from various EC MAST research projects as well as new findings worldwide. An improved approach to predicting the bed shear stresses due to waves and currents is included as well as a comprehensive section on the behaviour of mixtures of mud and sand, and sections on mathematical modelling and the role of intertidal processes. We have also included some case studies to demonstrate how the procedures can be applied to typical engineering studies. It is intended that this manual complements the earlier title *Dynamics of marine sands* (Soulsby, 1997), also published by Thomas Telford, which deals with the transportation by waves and currents of non-cohesive sand and gravel.

In writing this manual our intention has been to summarise a complex topic in a readily accessible manner. Where possible, we have first presented, in a summarised form, the current knowledge about a particular process and then presented procedures for making predictions. The data presented are intended to show the practising engineer which parameters are important in each of the processes and to enable broad estimates of the key processes to be made based on a limited knowledge of the field conditions. Some of the predictive approaches are readily applied by using the software tool SandCalc (version 2.00). Further information about SandCalc can be obtained from Wallingford Software at Howbery Park, Wallingford, Oxfordshire, OX10 8BA.

We hope that the manual is useful to those readers who are dealing with practical engineering problems as well as those researching the fundamental processes of cohesive sediment transport.

***Richard Whitehouse** obtained a degree in physical geography and a PhD on sediment transport in marine and aeolian environments. He joined HR Wallingford in 1988 where he has researched the behaviour of cohesive as well as non-cohesive sediments and applied the results to engineering studies. He currently works in the Marine Sediments Group on a range of research and consultancy projects.*

***Richard Soulsby** obtained degrees in physics and oceanography before specialising for 11 years in sediment transport research at the Institute of Oceanographic Sciences. Since 1985 he has worked on research and consultancy studies in this field at HR Wallingford, where he is currently head of the Marine Sediments Group.*

***Bill Roberts** obtained a degree in mathematical physics and a PhD in fluid dynamics at Edinburgh University and joined HR Wallingford in 1991. He currently works in the Ports and Estuaries Department, where he has specialised in the study of cohesive sediments, in both research and consultancy projects.*

***Helen Lowe (nee Mitchener)** obtained a degree in chemistry and oceanography and a masters degree in oceanography. She worked at Sir William Halcrow and Partners in water quality and coastal engineering from 1988 to 1990. She then moved to the Marine Sediments Group of HR Wallingford and spent nine years researching cohesive and mixed (mud and sand) sediment behaviour in the laboratory and the field. In 1999, she moved to DERA, where she currently works in the Underwater Sensors and Oceanography Department.*

# Acknowledgements

We are indebted to a panel of invited reviewers of the report version of this manual for their valuable comments and observations. The reviewers included Professor K. R. Dyer, Dr M. C. Christie, Mr I. S. Heijne, Dr K. J. Riddell, Professor D. Reeve and Mr P. B. Woodhead. Other valuable comments were made by our colleagues Dr M. P. Dearnaley, Dr C. A. Fletcher, who also contributed Section 2.7 on contaminants, and Ms L. J. Hall. We would also like to thank Mrs J. M. Clarkson and Mrs R. Smith for their adept typing of the manuscript and Ms L. J. Hall and Mr N. Reed who drafted the new figures.

Finally we also acknowledge the pioneering research on cohesive sediments carried out at HR Wallingford by Mr M. W. Owen and Mr M. F. C. Thorn, and Mr T. N. Burt, Dr E. A. Delo and Ms M. C. Ockenden who wrote the early versions of this manual from which we have drawn material.

This manual is published on behalf of the Department of the Environment, Transport and the Regions who part-funded its preparation. The views and information presented in the book are those of HR Wallingford and not necessarily those of the funding agencies.

HR Wallingford is an independent specialist research, consultancy, software and training organisation that has been serving the water and civil engineering industries worldwide for over 50 years in more than 60 countries. We aim to provide appropriate solutions for engineers and managers working in:

- water resources
- irrigation
- groundwater
- urban drainage
- rivers

- tidal waters
- ports and harbours
- coastal waters
- offshore.

**Address:** Howbery Park, Wallingford, Oxon, OX10 8BA, UK
**Internet:** http://www.hrwallingford.co.uk

# Notation

| | |
|---|---|
| $A$ | orbital amplitude of wave motion at the bed $= U_w T/2\pi$ (m) |
| $A_s$ | constant in Equation (5.3) |
| $B$ | constant in Equations (3.21), (5.14) and (5.15) |
| $c_1 \ldots c_5$ | consolidation coefficients in Equations (9.1) and (9.2) |
| $C$ | volume concentration of mud |
| $C_b$ | near-bed suspended sediment concentration by mass ($kg\,m^{-3}$) |
| $C_f$ | volume concentration of flocs in water |
| $C_i$ | concentration of suspended sediment in class $i$ ($kg\,m^{-3}$) |
| $C_{in}$ | internal concentration of grains inside a floc |
| $C_M$ | suspended sediment concentration or dry density (dry mass/volume) ($kg\,m^{-3}$) |
| $C_o$ | constant sediment concentration of mud layer ($kg\,m^{-3}$) |
| $C_s$ | saturated depth-averaged sediment concentration ($kg\,m^{-3}$) |
| $C(t)$ | suspended sediment concentration as a function of time ($kg\,m^{-3}$) |
| cos | cosine |
| $d_e$ | effective diameter of a floc (m) |
| $d_m$ | thickness of mud layer or depth of fluid mud (m) |
| $dm/dt$ | rate of change of mass on the bed per unit area ($kg\,m^{-2}\,s^{-1}$) |
| $d_{50}$ | median particle diameter (m) |
| $D_*$ | dimensionless grainsize or floc diameter |
| $e$ | 2·718281828 |
| $E1$, $E2$, $E3$, $E4$ | coefficients in Equations (4.1) and (4.2) |

| | |
|---|---|
| $f_w$ | wave friction factor |
| $f_{wr}$ | rough-bed wave friction factor |
| $f_{ws}$ | smooth-bed wave friction factor |
| $F$ | flocculation factor ($w_s/w_{sd}$) |
| $F_D$ | upward flux of density (sediment) (kg m$^{-2}$ s$^{-1}$) |
| g | acceleration due to gravity (m s$^{-2}$) |
| $h$ | water depth (m) |
| $H$ | height of water wave (m) |
| $H_{rms}$ | root-mean-square wave height (m) |
| $H_s$ | significant wave height (m) |
| $k$ | constant in Equation (5.4) |
| $k$ | wave number = $2\pi/L$ (m$^{-1}$) |
| $k_s$ | Nikuradse roughness length (m) |
| $K_I$ | (coefficient of) permeability (m s$^{-1}$) |
| $K_s$ | eddy diffusivity (m$^2$ s$^{-1}$) |
| $l$ | length scale (m) |
| ln | natural logarithm (to base e) |
| log | logarithm to base 10 |
| $L$ | wavelength of water wave (m) |
| $m$ | coefficient in Equations (5.4), (5.8) and (5.9) |
| $m$ | mass per unit area eroded (kg m$^{-2}$) |
| $m_e$ | erosion constant (kg N$^{-1}$s$^{-1}$ or m$^{-1}$ s) |
| $M$ | mass per unit area which is fluidised (kg m$^{-2}$) |
| $M_A$ | absolute moisture content (%) |
| $M_R$ | relative moisture content (%) |
| $N$ | coefficient in Equation (3.21) |
| $Q_m$ | sediment transport rate (kg m$^{-1}$ s$^{-1}$) |
| $Q_{smud}$ | transport rate of suspended mud (kg m$^{-1}$ s$^{-1}$) |
| $r$ | relative roughness in wave flows = $A/k_s$ |
| $R$ | tidal range (m) |
| $R_c$ | ratio of near-bed concentration to concentration near the water surface |
| $R_e$ | Reynolds number |
| $Ri_b$ | bulk Richardson number (Equation 6.5) |
| $Ri_f$ | flux Richardson number (Equation 5.2) |
| $Ri_g$ | gradient Richardson number (Equation 5.1) |
| $R_w$ | wave Reynolds number = $\frac{U_w A}{\nu}$ |
| $S$ | salinity |
| SC | equation contained in SandCalc software |
| sinh | hyperbolic sine |

| | |
|---|---|
| $t$ | time (s) |
| $T$ | wave period (s) |
| $T_m$ | mean period of water waves (s) |
| $T_n$ | scaling period for waves = $(h/g)^{0 \cdot 5}$ (s) |
| $T_p$ | peak period of wave spectrum (s) |
| $T_z$ | zero-crossing period of water waves (s) |
| $u_*$ | friction or shear velocity = $(\tau_0/\rho)$ (m $s^{-1}$) |
| $u_m$ | mean velocity of fluid mud layer (m $s^{-1}$) |
| $\bar{U}$ | depth-averaged current speed (m $s^{-1}$) |
| $U$ | current speed (m $s^{-1}$) |
| $U_w$ | wave orbital velocity amplitude just above the sea bed (m $s^{-1}$) |
| $U(z)$ | current speed at height $z$ above the bed (m $s^{-1}$) |
| $V_e$ | entrainment velocity for fluid mud (m $s^{-1}$) |
| $V_R$ | voids ratio |
| $w_{50}$ | median settling velocity (m $s^{-1}$) |
| $w_{50b}$ | median settling velocity immediately above the bed (m $s^{-1}$) |
| $w_f$ | settling velocity of flocculated sediment (m $s^{-1}$) |
| $w_s$ | settling velocity (m $s^{-1}$) |
| $w_{sd}$ | settling velocity of chemically dispersed sediment (m $s^{-1}$) |
| $w_{si}$ | settling velocity of sediment class $i$ (m $s^{-1}$) |
| $x$ | distance in direction of flow (m) |
| $z$ | height above sea bed (m) |
| $z_0$ | bed roughness length (m) |
| $\beta$ | angle of bed slope |
| $\dot{\gamma}$ | shear rate (rad $s^{-1}$) |
| $\gamma_d$ | unit weight of dry soil (N $m^{-3}$) |
| $\gamma_{sat}$ | unit weight of saturated soil (N $m^{-3}$) |
| $\gamma_{subm}$ | submerged unit weight of soil (N $m^{-3}$) |
| $\gamma_w$ | unit weight of water (N $m^{-3}$) |
| $\delta$ | boundary-layer thickness (m) |
| $\varepsilon$ | porosity |
| $\eta$ | water surface level relative to mean sea level (m) |
| $\eta_m$ | fluid mud-water interface level (m) |
| $\kappa$ | von Karman's constant = 0·40 |
| $\mu$ | dynamic viscosity (Ns $m^{-2}$) |
| $\nu$ | kinematic viscosity of water = $\mu/\rho$ ($m^2$ $s^{-1}$) |
| $\nu_B$ | bulk kinematic viscosity of suspension ($m^2$ $s^{-1}$) |
| $\pi$ | 3·141592654 |

| | |
|---|---|
| $\rho$ | density of fluid ($kg\,m^{-3}$) |
| $\rho_e$ | effective density of a floc (Equation 5.10) |
| $\rho_B$ | bulk density of sediment ($kg\,m^{-3}$) |
| $\rho_m$ | density of fluid mud layer ($kg\,m^{-3}$) |
| $\rho_s$ | density of sediment grains ($kg\,m^{-3}$) |
| $\sigma'_v$ | vertical effective stress ($N\,m^{-2}$) |
| $\tau$ | shear stress ($N\,m^{-2}$) |
| $\tau_b$ | shear-stress at bottom of fluid mud layer ($N\,m^{-2}$) |
| $\tau_c$ | current-only bed shear stress ($N\,m^{-2}$) |
| $\tau_{cr}$ | threshold shear-stress for motion of sand ($N\,m^{-2}$) |
| $\tau_d$ | critical shear-stress for deposition ($N\,m^{-2}$) |
| $\tau_{di}$ | critical shear-stress for deposition of sediment class *i* ($N\,m^{-2}$) |
| $\tau_e$ | threshold shear-stress for erosion ($N\,m^{-2}$) |
| $\tau_i$ | shear-stress at interface of fluid mud and water ($N\,m^{-2}$) |
| $\tau_0$ | applied bed shear-stress ($N\,m^{-2}$) |
| $\tau_m$ | mean bed shear-stress during a wave cycle under combined waves and currents ($N\,m^{-2}$) |
| $\tau_{max}$ | maximum bed shear-stress during a wave cycle under combined waves and currents ($N\,m^{-2}$) |
| $\tau_{rms}$ | root-mean-square bed shear-stress during a wave cycle under combined waves and currents ($N\,m^{-2}$) |
| $\tau_w$ | amplitude of oscillatory bed shear-stress due to waves ($N\,m^{-2}$) |
| $\tau(z)$ | shear-stress profile in the bed ($N\,m^{-2}$) |
| $\phi$ | angle between current direction and direction of wave travel |
| $\phi c_i$ | proportion of total concentration in sediment class *i* |
| $\omega$ | wave frequency $2\pi/T$ (rads $s^{-1}$) |

# Glossary

| | |
|---|---|
| Aerobic | The presence of free oxygen |
| Anaerobic | The absence of free oxygen |
| Algae | Plant primary organisms (like 'grass' of the oceans) |
| Bacteria/fungi | Special biological groups neither plant nor animal |
| Benthic | On bottom |
| Benthos | Organisms (plant and animal) that live on and in the sea bed |
| Biomass | Quantitative biological component |
| Biota | Biological component |
| Bioturbation | Mixing and disruption of sediment structure by living organisms (e.g. worms burrowing, bird footprints, tracks etc.) |
| Bulk density | Mass of mixture (solids and water) per unit volume of mud matrix |
| CEC | Cation Exchange Capacity – a measure of the flux exchange of ions in and out of muddy sediments (i.e. within the clay mineral structure) (defined in molecular equivalents per 100 g of dry sediment) |
| Chl 'a' | Mass of chlorophyll 'a' (photosynthetic pigment); an indicator of biological productivity |
| Colloidal Carbohydrate | Chemical polymeric compounds which can act as adhesives to further bind sediments secreted by micro-organisms, e.g. EPS (Extracellular Polymeric Substances) |

| | |
|---|---|
| Dry density | Mass of solids only per unit volume of mud matrix |
| Ecology | Interactions between living organisms and their environment |
| Fauna | Animal component of biology |
| Flora | Plant component of biology |
| Ligands and chelators | Molecules which form compounds/complexes with metal ions thereby influencing the behaviour of the metal |
| Littoral | Of the coast |
| LOI | Loss on Ignition (mass % by weight lost by ignition at 450 °C (INTRMUD Protocols, 1997) – indicator of carbon (organic) content, i.e. biomass |
| Micro | Size range 20–200 μm, e.g. microalgae |
| Mineral density | Mass of mineral solid (only) per unit volume of pure mineral |
| Neritic | On shore (continental shelf up to around 200 m depth) |
| Pelagic | In water column |
| pH | A term used to describe the indication of hydrogen ion concentration of dilute solutions, where the pH value is equal to $-\log 10[H^+]$. Solutions of pH between 0 to 7 are acidic, of pH 7 are neutral, and of pH between 7 to 14 are alkaline |
| Phyto (prefix) | Of plant kingdom |
| Reducing/oxidising | Conditions in sediment can be defined as either reducing or oxidising depending on the availability of oxygen; a reducing environment is one where free oxygen is deficient, while an oxidising environment is one where oxygen is readily available |
| Redox | The redox potential is the voltage difference at an inert electrode immersed in a reversible reducing/oxidising system and gives an indication of the state of oxidation of a system. Positive redox values are oxidising conditions while negative values are reducing conditions |

| | |
|---|---|
| RMS | Root-mean-square |
| SAR | Sodium Absorption Ratio – a measure of the relative abundance of $Na^+$ to the two other most common cations $Ca^{2+}$ and $Mg^{2+}$ |
| Zoo (prefix) | Of animal kingdom |
| % mud | Sediment particles <63 μm, by weight |
| % sand | Sediment particles >63 μm, by weight |

# Contents

# *Illustrations*

*Examples*



*Tables*

*Figures*

# Introduction

# 1. Introduction

## 1.1. GENERAL

The ability to predict the movement of cohesive sediment within coastal, estuarine or inland waters has a significant economical and ecological importance in the development of new engineering works and the maintenance of existing installations. The future viability of a proposed new port, for example, could depend largely on the cost of routine dredging necessary to sustain its accessibility to shipping. Many other schemes, such as the reclamation of intertidal flats, or the construction of flood protection structures or the laying of outfalls, also require a sound engineering appraisal of the likely changes in the patterns of sediment movement which will result after the scheme is built. Furthermore, the capability to predict the movement of cohesive sediment is crucial in the understanding of the distribution of certain pollutants, in particular heavy metals which are adsorbed on to clay and silt particles.

This manual was developed from, and replaces, the earlier HR Wallingford Report SR 309 *Estuarine muds manual* by Delo and Ockenden (1992) and its preceding versions (Delo and Burt, 1986, and Delo, 1988a). The present manual draws on a wide range of new data and new concepts in mud research, in addition to the material presented in the earlier reports and reviews by Dyer (1986), Owen (1977) and Van Rijn (1993). Other important reference works include those by Mehta and Dyer (1990) and USACE (1998), as well as Coussot (1997) on rheology and Terzaghi *et al.* (1996) on soil mechanics. Many of the valuable new research results obtained in the EC MAST collaborative projects G8M Coastal Morphodynamics (Stive *et al.*, 1995), INTRMUD (Dyer, 1998a) and COSINUS (e.g. Winterwerp *et al.*, 2000) have been included in this manual.

The manual is intended to be read as an adjunct to the book *Dynamics of marine sands* (Soulsby, 1997) or the equivalent HR Report SR466. To avoid a great deal of repetition, reference is made to information and equations in *Dynamics of marine sands*, particularly as regards the hydrodynamic quantities, which are largely similar for flow over sand or mud beds.

In common with *Dynamics of marine sands*, this manual is aimed at presenting mud research in a unified and easy-to-use fashion, without lengthy derivations and discussions. This manual summarises, in an engineering form, the main processes of cohesive sediment behaviour, namely, erosion, transport, deposition and consolidation. Sub-sections are given on *Knowledge* and *Procedure* where possible. The information presented under *Knowledge* is intended to show the practising engineer which parameters are important in each of the processes and, with the *Procedure* given, enable broad estimates of the rates of erosion, transport, deposition and consolidation to be made based on a limited knowledge of the field conditions. A wide range of units have been used in the literature and it has not always been possible to standardise units in the writing of this manual.

The behaviour of cohesive sediment varies considerably in quantitative terms from one location to another. Therefore, it is crucial for the engineer to appreciate that estimates based on the data presented herewith may well be in error by up to a factor 5.

For most serious engineering problems involving cohesive sediment it would be essential to undertake a detailed study. This would involve some of the following techniques: site visit with or without a programme of monitoring, laboratory testing of sediment, numerical modelling of hydrodynamics and sediment transport, and physical modelling of hydrodynamics.

## 1.2. SANDCALC SOFTWARE PACKAGE

To facilitate the quick and accurate calculation of many of the quantities featured in the manual, a software package SandCalc version 2.00 has been developed to complement the manual. This gives easy access to over 70 of the equations quoted in *Dynamics of marine sands*, plus 20 of the equations in this manual, in a Windows-based menu system for use on a PC.

The menu system is structured according to the same headings and sub-headings as are used in *Dynamics of marine sands* and this manual. Equations available in SandCalc are labelled 'SC' against the equation number in this manual, and the title of the equation as it appears in SandCalc is usually self-evident. Where necessary, some additional explanation of methods available through SandCalc are given in the appropriate place in the manual.

Quantities calculated by SandCalc as outputs from one equation are automatically carried as inputs to other equations. Input and output values are checked to ensure they fall within an allowable range, and default values are given for some parameters.

In version 2.00 of SandCalc, facilities are included to back-link to equations required to calculate input quantities, and to automatically work through procedures to solve some of the more commonly encountered problems.

Most of the worked examples and case studies given in the manual can be calculated easily using SandCalc. In the worked examples, the results of each step of the calculations are rounded to three significant figures. The accumulation of errors leads, in some cases, to small differences between the numbers in the example and the value given by SandCalc. In these cases, the SandCalc value is the more accurate one.

SandCalc version 2.00 is available from Wallingford Software, HR Wallingford Ltd, Howbery Park, Wallingford, Oxfordshire, OX10 8BA, UK. It can be purchased directly from their web-site: http://www.wallingfordsoftware.com.

# Sediment properties

# 2. Sediment properties

## 2.1. GENERAL

Muddy sediments occur commonly in estuaries, coastal embayments, and in areas of the continental shelf where both currents and waves exert only weak forces on the bed. In UK waters, the latter includes areas in the western and eastern Irish Sea, the northern North Sea, and the Firth of Clyde. The public perception is that mud is a dirty, sticky, dark-coloured and evil-smelling nuisance, whereas sand is an asset, conjuring up pictures of golden or silver holiday beaches. There is some justification for this prejudice. The stickiness is a defining characteristic of muds, which are technically classed as *cohesive* sediments. The dark colour and the smell arise from anaerobic decomposition of organic matter, which is often a major constituent of muds.

Whereas in sandy sediment it is the particle size of the bed sediment which controls the mobility and transport of sediment, with cohesive sediments it is the bulk properties of the admixture that determine the behaviour of the sediment. Generally speaking a sediment containing more than about 10% by mass of fine material (i.e. sieved material finer than 63 μm) may exhibit cohesive properties. A moist sample can be tested to see if it falls apart (non-cohesive) or clings together (cohesive).

Many different definitions have been adopted in the literature to express the bulk properties of a sediment bed. Some of these are summarised in Table 1 along with the conversion to the other commonly used parameters (see Table 2). Another parameter, not included in Table 2 but sometimes used for mud measurements, is the relative moisture content $M_R$, defined as the ratio of the mass of water to mass of solid particles. It is related to the dry density $C_M$ by the expression:

$$M_R(\%) = \left(\frac{\rho}{C_M} - \frac{\rho}{\rho_s}\right) \times 100 \qquad (2.1)$$

Given $M_R$ the following expression can be used to calculate the volume concentration $C$:

$$C = \left(1 + \frac{M_R \rho_s}{100\rho}\right)^{-1} \qquad (2.2)$$

In the above, $\rho$ is the density of water and $\rho_s$ the density of the sediment grains. Other measures of concentrations can be derived using the formulae in Table 2. The relationship between some of these parameters is shown graphically in Figure 1 for a typical

*Table 1. Bulk properties of mud beds and suspensions*

| Quantity | Usage | Definition | Symbol |
|---|---|---|---|
| Volume concentration<br>Packing density<br>Packing fraction | Suspensions (theory)<br>Bed (soil mechanics)<br>Bed (soil mechanics) | $\frac{\text{Volume of grains}}{\text{Volume of mixture}}$ | $C$ |
| Mass concentration<br>Dry density | Suspensions (experimental)<br>Bed (cohesive) | $\frac{\text{Mass of grains}}{\text{Volume of mixture}}$ | $C_M$ |
| Porosity | Bed (non-cohesive) | $\frac{\text{Volume of water}}{\text{Volume of mixture}}$ | $\varepsilon$ |
| Voids ratio | Bed (soil mechanics) | $\frac{\text{Volume of water}}{\text{Volume of grains}}$ | $V_R$ |
| Suspension density<br>Bulk density<br>Wet density | Suspensions (hydrodynamics)<br>Bed (non-cohesive)<br>Bed (cohesive) | $\frac{\text{Mass of mixture}}{\text{Volume of mixture}}$ | $\rho_B$ |
| Relative moisture content<br>Water content | Bed (cohesive)<br>Bed (soil mechanics) | $100 \times \frac{\text{Mass of water}}{\text{Mass of grains}}$ | $M_R$ |
| Absolute moisture content | Bed (cohesive) | $100 \times \frac{\text{Mass of water}}{\text{Mass of mixture}}$ | $M_A$ |

Note. In the context of mud, ‘grains’ is equivalent to dry mud, ‘water’ is equivalent to water in mud, and ‘mixture’ is equivalent to wet mud.

*Table 2. Conversions between measures for sediment-water mixtures*

| From / To | $C$ | $C_M$ | $\varepsilon$ | $V_R$ | $\rho_B$ |
|---|---|---|---|---|---|
| $C$ | – | $C_M/\rho_s$ | $1-\varepsilon$ | $\dfrac{1}{1+V_R}$ | $\dfrac{\rho_B-\rho}{\rho_s-\rho}$ |
| $C_M$ | $\rho_s C$ | – | $\rho_s(1-\varepsilon)$ | $\dfrac{\rho_s}{1+V_R}$ | $\dfrac{\rho_s(\rho_B-\rho)}{(\rho_s-\rho)}$ |
| $\varepsilon$ | $1-C$ | $1-C_M/\rho_s$ | – | $\dfrac{V_R}{1+V_R}$ | $\dfrac{\rho_s-\rho_B}{\rho_s-\rho}$ |
| $V_R$ | $\dfrac{1-C}{C}$ | $\dfrac{\rho_s-C_M}{C_M}$ | $\dfrac{\varepsilon}{1-\varepsilon}$ | – | $\dfrac{\rho_s-\rho_B}{\rho_B-\rho}$ |
| $\rho_B$ | $\rho(1-C)+\rho_s C$ | $\rho+C_M\left(\dfrac{\rho_s-\rho}{\rho_s}\right)$ | $\rho\varepsilon+\rho_s(1-\varepsilon)$ | $\dfrac{\rho V_R+\rho_s}{1+V_R}$ | – |

$\rho$ = density of water (typically 1000 kg m$^{-3}$ for fresh water, 1027 kg m$^{-3}$ for sea water).
$\rho_s$ = density of grain material (typically 2650 kg m$^{-3}$ for quartz, see Table 6 for common clay minerals). See Table 1 for meaning of symbols.
Example: to convert from porosity to mass concentration $C_M = \rho_s(1-\varepsilon)$.

combination of mud and water parameters. This figure can be used to make an approximate conversion between parameters.

Other measures which may be encountered are the unit weight of dry soil, $\gamma_d$ and the unit weight of soil in a saturated state $\gamma_{sat}$, usually quoted in kN m$^{-3}$. These are defined by:

$$\gamma_d = gC_M \tag{2.3}$$

$$\gamma_{sat} = g\rho_B \tag{2.4}$$

where g = 9·81 m s$^{-2}$.
For example, a sediment with bulk density $\rho_B$ = 1800 kg m$^{-3}$ is equivalent to $\gamma_{sat}$ = 17·7 kN m$^{-3}$.

Site investigation reports often contain reference to the submerged unit weight $\gamma_{subm}$ of the soil (kN m$^{-3}$). This is defined as:

$$\gamma_{subm} = \gamma_{sat} - \gamma_w \tag{2.5}$$

where $\gamma_w$ is the unit weight of water, i.e. $\rho g$.

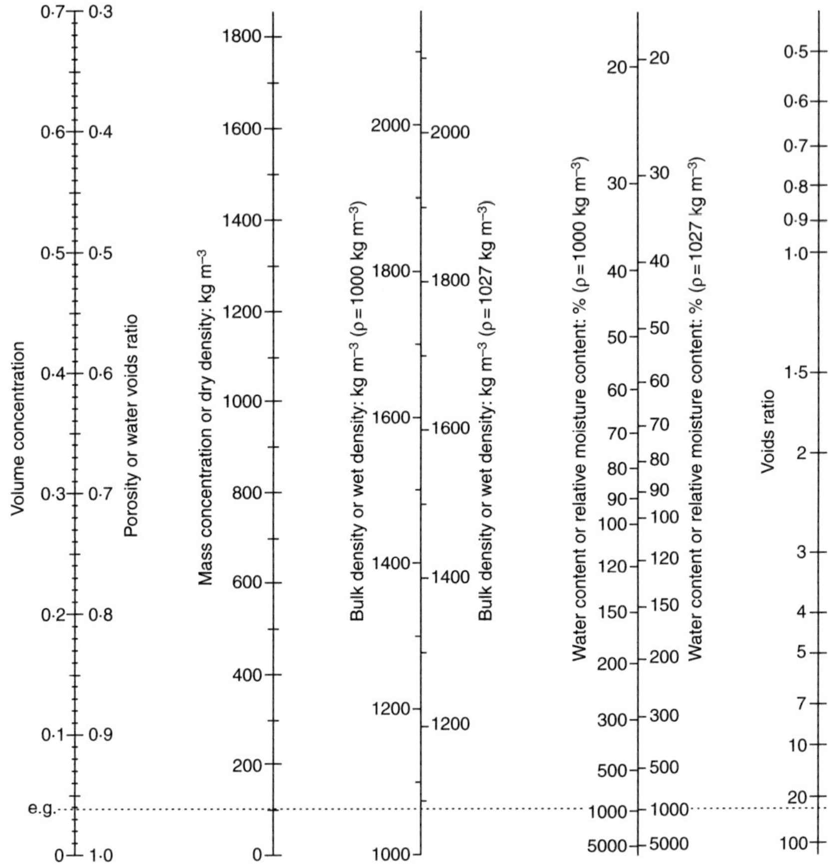


*Figure 1. Conversions between commonly used measures for sediment beds*

## 2.2. DIFFERENT STATES OF MUD

The processes of deposition, consolidation and erosion of cohesive sediment are controlled by a complex array of physical, chemical and biological factors, which are only partly understood. Any attempt to predict the movement of cohesive sediment must first investigate the nature of the hydrodynamics of the water and then relate the movement of water to the movement of cohesive sediment. As yet, it is not possible to predict the behaviour of a cohesive sediment from its physical and chemical properties alone and the principal thrust of research has been to determine in the laboratory, for a given set of flow conditions, the behaviour of a sample of the cohesive sediment taken from the field. Solutions in this instance, are therefore, based on empirical data and have limited value to other sites.

The complexity of cohesive sediment may be demonstrated by reference to a characterisation of cohesive sediment (Hayter and Mehta, 1982) reproduced in Table 3. The number of parameters which need to be determined to completely describe a cohesive sediment is quite considerable. Hence it is easy to understand why studies of cohesive sediment have been empirical and site specific rather than of a more fundamental nature. Fortunately, not all these parameters are needed in order to make predictions of the most commonly needed aspects of mud behaviour, as described later.

Most of the laboratory studies to date, however, have had the drawback of considering only one process in isolation, e.g. deposition or erosion, and even then usually at a constant rate of flow. In natural conditions the processes are often strongly cyclic with the deposition, partial consolidation and re-erosion of cohesive sediment occurring repeatedly with the tides. Laboratory simulation of the tidal cycle in relation to the physical processes of cohesive sediment has been undertaken at HR Wallingford (Delo, 1987) and some of the results are reported herein.

Estuarine cohesive sediment, commonly called mud, is composed primarily of silt and clay. The size distributions of five muds investigated at HR Wallingford are given in Figure 2 – expressed in terms of their Stokes equivalent diameter, determined by sedimentation of a dispersed sample in the laboratory. The mineralogy and cation exchange capacity of three of the muds are presented in Table 4. Mud contains a large proportion of very small particles which have a large specific area such that the effect of the surface physico-chemical forces becomes as important as the effect of gravity forces. Some of these individual particles are less than 1 micron in diameter and may be kept in suspension by Brownian motion alone. Flocculation of particles will take place when the net physico-chemical interparticle forces become attractive.

Cohesive sediment can be considered to exist in four states. The processes linking these four states are indicated schematically in Figure 3 and may be described as a mobile suspended

*Table 3. Characterisation of sediment (after Hayter and Mehta, 1982)*

| | |
|---|---|
| 1. | Type of material<br>(*a*) Clay minerals<br>(i) Clay mineral alone<br>(ii) Mixture of clay minerals in varying proportions<br>(iii) Mixture of clay mineral and non-clay mineral, both in the fine sediment range<br>(*b*) Soils, muds and clay material<br>(i) Mixture of cohesive and non-cohesive (such as sand) sediments<br>(ii) Mixture of clay material and organic matter or organic compounds<br>(iii) Sediments from natural environment (unclassified)<br>(iv) Sediments from natural environment (classified according to Soil Classification System)<br>(*c*) Non-sediment fine materials |
| 2. | Nature of clay structure<br>(*a*) Electrical forces acting between particles<br>(i) Net energy of attraction<br>(ii) Double layer thickness<br>(*b*) Particle arrangement or fabric consisting of texture and particle orientation |
| 3. | Particle size distribution<br>(*a*) Median diameter<br>(*b*) Effective size<br>(*c*) Uniformity coefficient<br>(*d*) Curvature coefficient |
| 4. | Cation Exchange Capacity |
| 5. | Exchangeable sodium percentage |
| 6. | Sodium Adsorption Ratio of clay |
| 7. | Dielectric constant |
| 8. | Silica-sesquioxide ratio |
| 9. | Chemical composition |
| 10. | Specific Gravity |
| 11. | Hydration or adsorbed water |
| 12. | Antecedent water |
| 13. | Aging |
| 14. | Colour |
| 15. | Odour |
| 16. | Texture |
| 17. | Grain properties |
| 18. | Biological matter |

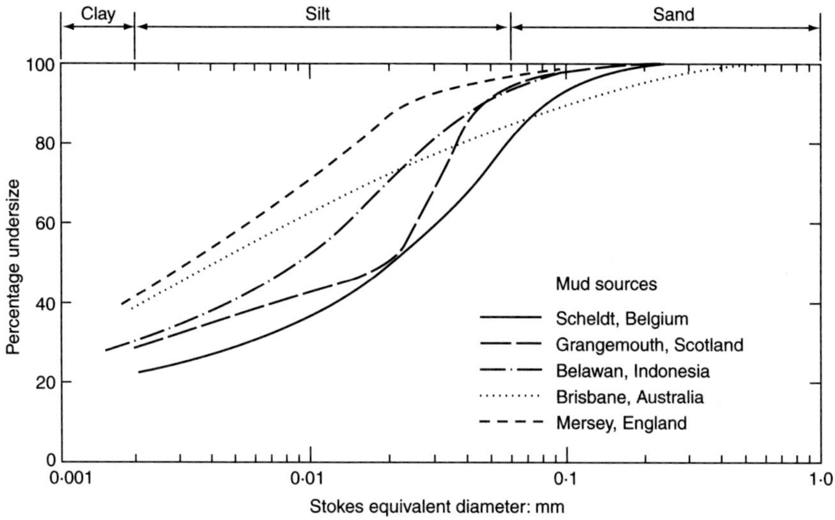


*Figure 2. Particle size distributions of some muds*

sediment, a high concentration near bed layer which is sometimes referred to as fluid mud, a newly deposited or partially consolidated bed, and a settled or consolidated bed. This classification is elaborated on in Section 5.2.

The four processes of cohesive sediment of primary interest to the engineer are erosion and transport, deposition and

*Table 4. Mineralogy and cation exchange of three muds tested at HR Wallingford*

| Mud source | Grangemouth | Brisbane | Belawan |
|---|---|---|---|
| Cation Exchange Capacity: meq/100 g | 20 | 35 | 25 |
| % clay minerals | 51 | 50 | 75–80 |
| Montmorillonite | nil | 30 | 15–20 |
| Kaolinite | 17 | 15 | 30 |
| Illite/mica | 17 | 5 | 30 |
| Chlorite | 17 | nil | trace |
| % non-clay minerals | 39 | 50 | 20 |
| % organics and minor constituents | 10 | nil | traces |

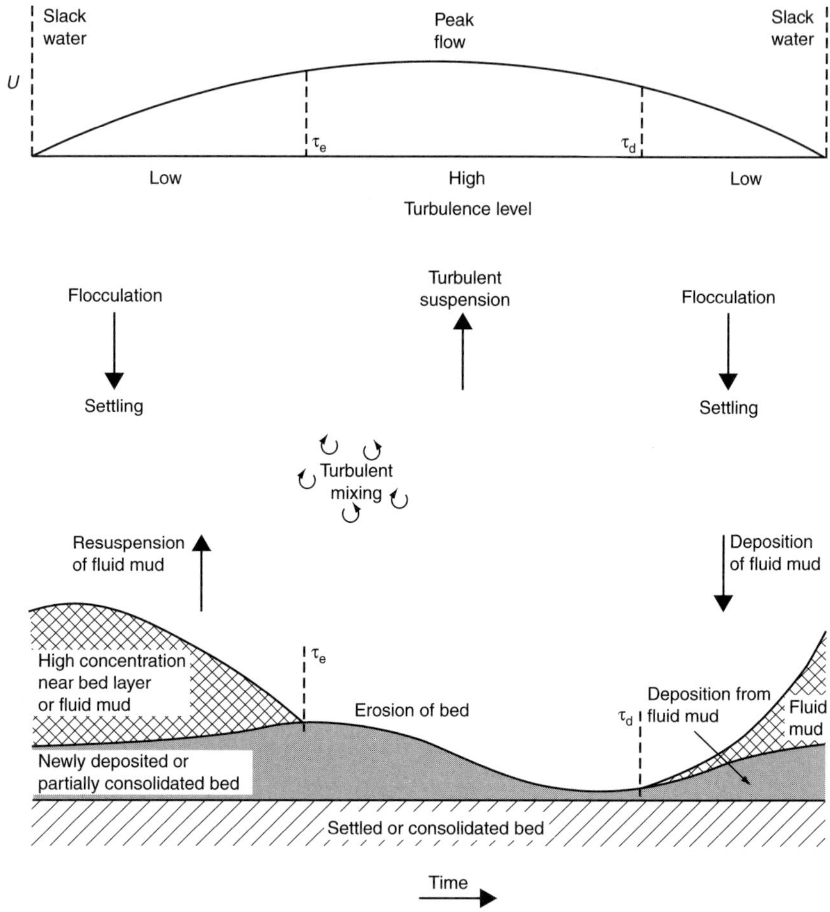


*Figure 3. States of cohesive sediments*

consolidation. *Erosion* is the removal of sediment from the surface of the bed due to the stress of the moving water above the bed. *Transport* is the movement of suspended mud and high concentration layers on or near the bed by the flow. *Deposition* involves the settling through the water column and on to the bed of flocculated sediment. *Consolidation* of a deposit is the gradual expulsion of interstitial water by the self weight of the sediment accompanied by an increase in both the density of the bed and its strength with time.

## 2.3. DESCRIPTION OF NATURAL SEDIMENTS

Natural sediments rarely consist of only either mud or sand and the majority of estuarine sediments are comprised of a combination of grain sizes including sands, silts and clays, and even gravels. The mineralogical make-up is also varied as sand is primarily quartz but clay particles are comprised of metal silicates in different chemical forms (Section 2.4.2). Other sediment constituents may include organic particles and polymers, oil, and shell fragments.

The fabric of the sediment becomes important when considering sediment mixtures because these sediments are commonly deposited in layers (Torfs *et al.* 1996), with a degree of vertical sorting. This means that a natural undisturbed sediment may be comprised of layers of mud and sand and thus its sediment transport properties will reflect its structural composition. Furthermore, the sediment may then undergo mechanical and biological reworking which will break up the vertical structure, as well as compaction, consolidation, erosion and further deposition. Mixed sediments are thus classified either as a homogeneous or a layered (heterogeneous) sediment and are in practice a complex, three-dimensional combination of textural and sedimentological constituents.

Historically, sediments have been treated as either muds or sands because the characteristics and resultant behaviour are very different. As a result there are well-established sediment transport formulas for non-cohesive sediments (Soulsby, 1997), and some for muddy sediments, as will be described later, but few parameterisations exist for true sediment mixtures. Most references to mixed sediment behaviour in the literature are incidental and describe the effect of adding sand to mud or vice versa, but still treating the sediment as if it were a mud or a sand.

An assessment of a mixed sediment should include both sediment composition, and the texture of the sediment matrix. It is recommended to collect vertical core samples of the sediment and assess the degree of sediment sorting and layering by detailed observation or X-ray photography. A grain size analysis can then be undertaken to determine the amount of mud and sand in the mixture. Sediment can then be treated as a mud or sand based on the relative amounts of the constituents. In practice, however,

*Table 5. Physical properties*

| Inherent properties |
|---|
| $\rho_b$ bulk density ($kg\,m^{-3}$) |
| $C_M$ dry density ($kg\,m^{-3}$) |
| $\rho_s$ mineral (grain) density ($kg\,m^{-3}$) |
| Water content/voids ratio |
| Vertical density profile |
| Vertical structure (layering) |
| Grain-size distribution |
| $d_{50}$ median grain-size (m) |
| Sand content (% greater than 63 μm by weight) |
| Mud content (% less than 63 μm by weight) |
| Mineralogical composition |
| Organic content (% by weight by loss on ignition) |
| pH |
| Temperature (degrees Celsius) |
| Pore water salinity |
| Redox potential (volts) |
| CEC Cation Exchange Capacity (meq/100g) |
| SAR Sodium Adsorption Ratio |
| Rheological parameters (e.g. yield stress, $N\,m^{-2}$) |
| Chlorophyll 'a' content (μg/g) |
| Carbohydrate content (mg/g) |
| **Sediment behaviour** |
| *Erosion* |
| $\tau_{cr}$ critical shear-stress for threshold of erosion at surface ($N\,m^{-2}$) |
| $m_e$ erosion constant ($kg\,N^{-1}\,s^{-1}$ or $m^{-1}\,s$) |
| $dm/dt$ erosion rate ($kg\,m^{-2}\,s^{-1}$) |
| $z_0$ roughness length (m) |
| Mode of erosion (cohesive, transitional, non-cohesive) |
| *Deposition* |
| $\tau_d$ critical shear-stress for deposition of sediment ($N\,m^{-2}$) |
| $w_{50}$ median settling velocity ($m\,s^{-1}$) of a flocculated sediment |
| $dm/dt$ deposition rate ($kg\,m^{-2}\,s^{-1}$) |
| *Consolidation* |
| Gel point ($kg\,m^{-3}$) |
| $dh/dt$ consolidation rate ($m\,s^{-1}$) |
| Density profiles as a function of time and depth |
| $\sigma'_v$ vertical effective stress ($N\,m^{-2}$) |
| $K_I$ permeability ($m\,s^{-1}$) |

only a relatively small amount of cohesive material, of about 10% dry mass by weight is required to convert a sandy bed into one exhibiting cohesive properties and then this sediment should be treated as if it were a mud.

Natural aquatic sediments are likely to consist of a mixture of mud/sand/gravel in varying proportion and will have a specific three-dimensional structure. Real sediment properties will result from a combination of physical, chemical, biological and environmental conditions. The physical properties of the material can be split into two groups (Table 5). Firstly, properties which are *inherent* properties of the sediment (e.g. mineralogy, density, grain size distribution, water content) and, secondly, the *behavioural* properties (i.e. how it erodes, deposits and consolidates under different environmental conditions). Mitchener (1996) and Berlamont *et al.* (1993) give examples of methods and instrumentation that can be used to measure these sediment properties.

## 2.4. BIOLOGICAL AND CHEMICAL PROCESSES

Marine and estuarine environments are ecologically active zones, and, as a result, natural sediments are comprised of biological and chemical components as well as the more inert mineralogical grains. In practice, sediments are complex combinations of different granulometry and structural make-up, as well as significant components of biota (living and dead) within and on top of benthic sediments. The chemical environment of bottom sediments can also affect their properties and behaviour. It is therefore unrealistic to treat sediment as grains only and to obtain a true picture of natural sediments they are best treated within their own 'microenvironment' of biological and chemical interactive processes. Cohesive sediment is particularly affected by chemical and biological environment as the binding force between clay minerals in muddy sediments is primarily of a chemical electrostatic nature and is therefore influenced by small-scale biochemical changes. The biological and chemical environments of sediments can influence all aspects of their physical behaviour – deposition (Ginsberg and Lowenstam, 1958), sediment fabric and bottom features, and erodibility (Heinzelmann and Wallisch, 1991).

### 2.4.1. *Biological processes*

The benthic community can constitute a significant proportion of the sediment mass, typically over 200 g dry weight of large animals per $m^2$ of surface sediments in estuarine and continental shelf regimes (Meadows and Campbell, 1978). Their presence and activity can affect the sediment in a number of different ways; biostabilisation, biodestabilisation, bioturbation, and micro-topography.

#### *Biostabilisation*

Some biological processes act to stabilise the sediment, which has been well-documented in the literature. Paterson (1994) gives a good review of the major mechanisms of biogenic stabilisation and there are also good papers on the subject by Frostick and McCave (1979), Grant and Gust (1987) and Dade *et al.* (1990). Even as early as the mid-nineteenth century, Huxley (1868) observed a pervasive slime or mucilage on the ocean floor, and recognised its importance on sediment stabilisation.

An important process of biological stabilisation is the secretion of adhesive mucous compounds called extracellular polymeric substances (EPS). These substances are secreted by microalgae (diatoms and blue-green algae), bacteria, fungi and microfauna. Some worms, molluscs, crustaceans and other sediment fauna also excrete mucous, which acts to bind the sediment grains together. In addition, the formation of biological network structures and mats create a network cage around sediment fabric and provide additional strength to resist matrix break-up. This can vary from the large scale, such as mangrove roots in tropical swamps, to microscopic fungi and bacteria threads which act to bind individual grains together. Some animal burrows, particularly those forming armoured tubes within the sediment can increase the strength of the sediment (Meadows and Meadows, 1991). Paterson (1994), quoted the possible effect of biogenic stabilisation on sediment motion as being between 100% and 900% increase for non-cohesive sediment, and 45% to 500% increase in stabilisation for cohesive sediment.

#### *Biodestabilisation*

Alternatively, but to a lesser extent in general, biological activity can result in the sediment being more susceptible to erosion.

The activity of digging and burrowing can weaken the sediment structure and create zones of weakness around which erosion will preferentially occur (Eckman *et al.*, 1981). The ingestion of sediment grains and production of faecal pellets changes the surface make-up of the water-sediment boundary. Filter feeding animals can segregate the grains within the locality of the burrow and grazing of surface organic food matter can also change the surface sediment characteristics (Young and Southard, 1978).

### *Bioturbation*

Bioturbation is the general term given to the reworking of the sediment by biological activity. Sediment is often deposited in layers, which reflect depositional events, and subsequent biological activity has the action of mixing the sediment, and introducing oxygen into otherwise anoxic areas. Bioturbation also changes the matrix structure of the sediment and results in surface formations at the sediment-water interface.

### *Microtopography and bottom roughness*

In terms of what the overlying water mass 'sees', the biological activity of benthic sediments can have pronounced effects on the bottom topography. This activity can result in the production of animal tracks, burrow and tube entrances, faecal trails and casts and shell debris which locally affect the flow regime above the bed surface (Mitchener *et al.*, 1996a, Heinzelmann and Wallisch, 1991). Tracks, burrows and tube worm protrusions into the flow and the flora and fauna itself affect the bottom roughness and change locally the hydrodynamic forces that the sediment surface experiences. Within the INTRMUD project work was undertaken to model the influence of the shells of the small snail *Hydrobia ulvae* and their tracks on the bed roughness (Sauriau *et al.*, 1999). Microbial mats, films and faecal deposits can smooth the bed surface and decrease the hydraulic stress experienced by the exposed sediment surface. Biological communities also move around in response to tidal forcing, light levels and diurnal processes and it is likely that the roughness of natural sediment will vary in space and time depending on the prevailing ecological processes.

### 2.4.2. Chemical processes

#### Chemical environment and mineralogy

Sand grains are commonly comprised of quartz minerals such as chert and flint, which are chemically hard and stable. The clay mineral groups are the feldspars, micas and aluminosilicates. These minerals possess inherent ionic charges and can undergo chemical exchanges of their elemental make-up by processes such as weathering, and local solute environments. Cohesive sediments are primarily comprised of four clay mineral types:

- kaolinite
- illite
- chlorite
- montmorillonite.

Table 6 lists the characteristics of these clay minerals.

They are termed metal silicates and their chemical structure is that of layers of silica tetrahedra and aluminium hydroxide joined loosely by metal ions and water molecules. They are thus easily broken up by natural degradation processes acting along these cleavage planes and therefore typically occur as plate-like fragments in sediments almost exclusively in the size fraction less than 2 μm (Dyer, 1986).

The reason why sediments containing significant proportions of clay minerals are cohesive is largely attributable to the chemical forces acting between the grains on a small scale. The grains are characterised by a plate-like appearance, with negatively charged faces and positively charged edges. They can thus become stuck together in a 'card-house' structure during

*Table 6. Characteristics of common clay minerals*

| | CEC: meq/g | Mineral density: kg m$^{-3}$ |
|---|---|---|
| Kaolinite | 0·03–0·1 | 2600 |
| Illite | 0·2–0·3 | 2660–2720 |
| Chlorite | 0·2–0·3 | 2600–3000 |
| Montmorillonite | 0·8–1·2 | 2500–2800 |

Source: Terzaghi *et al.* (1996)

flocculation by the electrostatic forces at the edge/faces. However, the natural situation is more complex than this, and the surrounding water environment and ionic constituents affect the balance of chemical species both within and surrounding the mineral particles. Typically in saline water free cations such as sodium $Na^+$, potassium $K^+$, magnesium, $Mg^{2+}$ and calcium $Ca^{2+}$ form a layer around the grains on the negatively-charged plate faces, and anions such as hydroxide $OH^-$ collect around the positively-charged plate edges. The overall effect is the formation of an electrical double-layer around the clay particle. The combination of a molecular attractive force and the electrostatic forces between neighbouring clay grains leads to bonding between clay particles and gives rise to its cohesive or 'sticky' nature.

The microchemical environment surrounding the clay grains, i.e. the interstitial pore water, plays an important part in the chemical interactions between the grains. In freshwater environments there are few ions in the water to form a double layer around the clay plates and the electrostatic forces dominate. This results in weak bonding, poor flocculation and aggregates which break up readily under turbulent shearing. In brackish and saltwater, though, there is increased ionic concentration and the double layer is important and flocculation, or particle bonding, increases with salinity up to about 2–3 ppt. Organic materials, such as the polymeric substances described earlier (Section 2.4.1), are also adsorbed preferentially on to the clay particles and these also act to increase cohesion. Other chemical pore-water conditions also affect fine-grained sediments, namely temperature and chlorinity (or salinity), redox potential (a measure of the oxidation state), pH and the sodium adsorption ratio (SAR – defining the relative concentration of $Na^+$, $Ca^{2+}$ and $Mg^{2+}$).

## 2.5. FLOCCULATION AND SETTLING

The flocculation of sediment particles arises as a direct consequence of particles sticking together as they are brought into contact with each other. Collision and cohesion are therefore the essential processes of flocculation. These factors are virtually independent of one another and are well described elsewhere by Krone (1962), Partheniades (1962) and Van Leussen (1988).

Cohesion is governed by the electrostatic forces described in Section 2.4.2. The action of these forces is strong at short distances, but falls inversely with the seventh power of distance for spheres and inversely with the square or cube of the distance for parallel plates. Particles will cohere if these short-range forces dominate over the repulsive forces generated by the clouds of cations around the particles. The strength of the repulsive forces depends on the charge on the mineral surface, which is determined by the mineral composition, and by the amount and types of cations present in the suspending fluid.

Collisions of particles are the result of one of three mechanisms, namely, Brownian motion of the suspended particles, internal shear of the water, and differential settling velocities of the particles or flocs. All three of these mechanisms operate in an estuary although it was postulated by Krone (1972) that the formation of large aggregates is predominantly due to internal shearing.

Nevertheless, the size of flocs formed by collisions from any of these mechanisms is limited by the maximum rate of internal shear that the flocs can withstand. It is evident, therefore, that internal shearing can both promote the growth of flocs and limit their size. Hence, suspended flocs should attain a maximum size given constant conditions of internal shear.

The size and settling velocity of the flocs may be much larger than that of the individual particles and rapid deposition may occur as a result of flocculation. The importance of flocculation may be demonstrated by considering the relationship shown in Figure 4 (Migniot, 1968). The flocculation factor, $F$, is the ratio of the settling velocity of flocculated sediment $w_f$ to that of the chemically dispersed sediment $w_{sd}$. This factor is seen to vary with mean particle diameter from a figure in the order of $10^4$ for particles with a diameter of 0·1 microns to unity for particles with a diameter of about 60 microns (0·06 mm). This implies that the cohesive behaviour of sediment ceases for particles with a mean diameter greater than about 60 microns.

The maximum floc size is governed by the particle size, concentration, mineralogy, pH and ionic strength of the mud. It is also controlled by the chemical composition of the pore water and suspending water, and by the hydrodynamic parameters of the

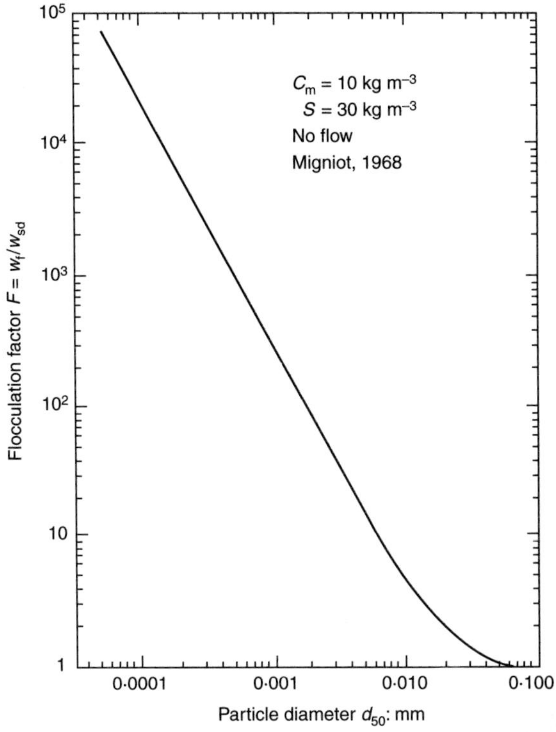


*Figure 4. Flocculation factor, F, as a function of median particle diameter*

water such as the velocity and turbulence structure, internal shear and bed shear stress. The main settling unit is therefore the floc rather than discrete particle grains as is the case for non-cohesive sediments. This dependence therefore inhibits the development of a set of universal equations.

These processes tend to remove one of the complicating factors associated with muddy sediments, that is while many sand particles can be approximated as spheres clay minerals are platy in nature. However, the processes of erosion, transport and deposition lead generally to agglomerations of particles which will be rounded naturally by hydrodynamic processes. It is only in the fluvial environment (where cohesion is usually negligible) that clay particles behave as individuals and the erosion and settlement processes will be largely determined by the size and shape of the individual particles.

## 2.6. SOIL MECHANICS AND RHEOLOGY

Soil mechanics is concerned with characterising the constituents of a sedimentary material (soil or consolidated bed) and the bulk behaviour of that material in terms of its resistance to loading. One of the standard texts on soil mechanics is Terzaghi *et al.* (1996). Rheology is concerned with determining the stresses and strains in a sediment-water mixture under the influence of some external loading (shear). The rheology of sediment-water mixtures applicable to marine sediment transport is discussed by Berlamont *et al.* (1993) and a more comprehensive text on the rheology of mud flows is Coussot (1997). Methodologies based on both soil mechanics and rheology have been applied to marine soils to assess their stability under foundation loads and the stability under earthquake and wave loading. Rheological investigations have also been used to examine the stability of the side slopes to navigation channels where the sediment can move under gravity (creep) or flow under gravity having first been liquefied by wave action (flow).

There are a range of parameters used in the hydraulic engineering, soil mechanics and rheological literature to define the bulk properties of a sediment (see Section 2.1, Table 1 and Table 3). A nomograph for converting between some of the soil bulk properties for typical conditions is presented in Figure 1.

### 2.6.1. *Soil mechanics*

#### *Knowledge*

In the soil mechanics literature there is a need to provide workable definitions of the complete spectrum of soil types from pure cohesionless sands to clays. Therefore a variety of soil parameter definitions have been prescribed and test procedures defined to help the engineer assess the performance of foundations (e.g. Terzaghi *et al.*, 1996). These measures of soil properties have also been used in an attempt to produce correlations between the hydraulic behaviour of the soil (i.e. threshold for erosion and erosion rate) and the properties of the bed such as bulk density and shear strength. However, the analogy between the force required to strip the surface layers of a soil through the action of flowing water and the force required to shear the bed using a shear vane (which measures the torque force required to rotate a vane of a given shape and size) is not that clear. Individual studies have demonstrated that a reasonable correlation exists between the two measures (e.g. Kamphuis and Hall, 1983). They found that the value of shear stress (and velocity) required for erosion increased with unconfined compressive strength, vane shear strength, plasticity index, and clay content of the soil, and with increasing consolidation pressure. The results of tests on a consolidated cohesive sediment (clay) showed the variation of threshold shear-stress for erosion with the first two of these parameters was linear. However, a universal relation of this type for both partially consolidated and consolidated sediments has not been achieved. One reason for this is that tests are carried out with different types of equipment, e.g. the shear vane integrates over a vertical element of the soil (>several mm and usually several cm at a depth of several cm in the bed) and the reading is related to the macroscopic properties of the soil; also different shear vanes will have different dimensions and operating procedures. Another reason is the variation in the methods used to prepare the test soils.

Despite these shortcomings the use of a vane shear device provides useful information on the spatial uniformity of the soil strength at a site. A measure of the temporal variation in resistance to erosion is possible if repeat measurements are taken in the same location over a period of days or throughout the year.

There are also a variety of devices available for making measurements of soil erodibility in-situ (Black and Paterson, 1997) which gives scope for determining the erosion properties directly. Using appropriate equipment a site-specific relation between shear strength or other bulk sediment properties and the erosion shear-stress and erosion rate in-situ can be derived.

## *Procedure*

The following approach can be taken to characterise the sediment.

1. Take field samples of the bed material for analysis of inherent properties (Table 5) in the laboratory.

2. Use a shear vane to determine the strength of the bed in the field and how it varies within the area of interest.

3. Use appropriate in-situ erosion equipment to determine the erosion threshold and erosion rate of the soil.

4. If required, determine site specific relationships between soil bulk properties and measures of erodibility.

## 2.6.2. Rheology

### Knowledge

The rheological characteristics of a cohesive sediment will influence how a sediment suspension modifies and interacts with the flow near the sea bed. They will also determine whether mud deposited on a slope can flow under gravity and how a mud bed responds to the action of waves. The rheological behaviour of cohesive sediments is strongly influenced by the concentration of sediment and physico-chemical factors such as salinity, pH, mineralogical composition, solid particle properties and organic matter. This influence is largely reflected in the way in which these factors affect the flocculation process (Bryant *et al.*, 1980; James and Williams, 1982; Verreet and Berlamont, 1988).

A number of different rheological models have been used to describe cohesive sediment. These models, which describe the sediments response to fluid loading, are given in Table 7, with a schematic representation of the flow curves given in Figure 5 (after Verreet and Berlamont, 1988).

Various methods have been devised for examining the rheology of mud-water suspensions (MAST G6M, 1993) and these can produce different results. For example, a controlled stress rheometer (the preferred method – MAST G6M, 1993) differs from a controlled shear rate instrument in that a shear stress is applied to the sample to be measured. This enables more meaningful low shear data to be produced since the instrument reacts in 'sympathy' to the fluid behaviour and does not force the material to move as is common in a controlled shear rate apparatus. Only equilibrium flow curves are capable of being analysed in a meaningful manner for time dependent non-Newtonian materials. For low concentrations it is generally recognised that if careful

*Table 7. Rheological models (after Verreet and Berlamont, 1988 – also see Figure 5). Reproduced by permission. From* Encyclopaedia of fluid mechanics, *© 1988, Gulf Publishing Company, Houston, Texas. All rights reserved*

| Model | Curve | Equations | Material type |
|---|---|---|---|
| Newtonian | 1 | $\tau = \eta \frac{du}{dz}$ | Dilute suspensions |
| | | $\eta_r = 1 + 2{\cdot}5\phi$ | Spherical particles |
| | | $\eta_r = \exp[2{\cdot}5\phi/(1 - k_1\phi)]$ | |
| Shear thinning (Pseudoplastic) | 2 | $\tau = m\left(\frac{du}{dz}\right)^n \quad n < 1$ | Flocculated clay slurries |
| | | $\eta_a = m\left(\frac{du}{dz}\right)^{n-1}$ | |
| Generalised Bingham | 3 | $\tau = \tau_0 + \eta_B\left(\frac{du}{dz}\right)^n$ if $\lvert\tau\rvert \geqslant \tau_0$; with $\frac{du}{dz} = 0$ if $\lvert\tau\rvert < \tau_0$ | |
| Bingham plastic | 3(a) | $\tau = \tau_{yB} + \eta \frac{du}{dz}$ | |
| Shear thickening (Dilatant) | 4 | $\tau = k\left(\frac{du}{dz}\right)^n \quad n > 1$ | Concentrated deflocculated clay slurries |
| Visco-elastic (Voigt solid, Kelvin body) | | $\tau = G\gamma + \mu\dot{\gamma}$ | |

$\tau$ = shear stress

$\eta$ = dynamic viscosity

$\eta_0$ = dynamic viscosity of medium

$\eta_r = \frac{\eta}{\eta_s}$ = relative viscosity

$\eta_d = \frac{d\tau}{d\dot{\gamma}}$ = differential viscosity

$\eta_a = \frac{\tau}{\dot{\gamma}}$ = apparent viscosity

$m$ = 'pseudoplastic viscosity'

$\tau_0 = \tau$ at $\dot{\gamma} = 0$

$\tau_{yB}$ = true Bingham plastic yield stress

$\tau_y$ = yield stress

$\tau_B$ = 'Bingham' yield stress

$\gamma$ = shear deformation

$\dot{\gamma}$ = shear rate

$\phi$ = solids volume fraction

$n$ = flow index

$k$ = shear-thickening analog of $m$

$k_1$ = constant

$G$ = rigidity modulus or shear modulus

experiments are carried out at very low shear rates then whatever the stress applied the material will flow (no yield stress) but it is all a matter of time scale (Bryant *et al.*, 1980; Jones and Golden, 1990). For a particular shear rate the shear stress increases with increase in concentration (see Figure 6). However, the concept of yield stress is dependent on the time of the experiment. In certain flow situations the existence of a yield stress can prove to be very useful, for correlations with various physical properties. Several researchers have published studies of direct yield measurements (James *et al.*, 1987). Relationships between rigidity modulus, yield stress and volume fraction of cohesive sediment and concentrated suspensions of Na-illite (a model sediment) have been published by Williams and Williams (1989). Migniot (1968) found a relation between the velocity for erosion and the yield stress of the soil (initial rigidity) determined in a rotating cylinder viscometer.

Most mud type suspensions exhibit Newtonian type behaviour at low concentrations but show non-Newtonian shear-thinning or thixotropic viscoelastic behaviour at higher concentrations. The transition from shear-thinning to viscoelastic properties takes place over a range of only a few percent of the solids

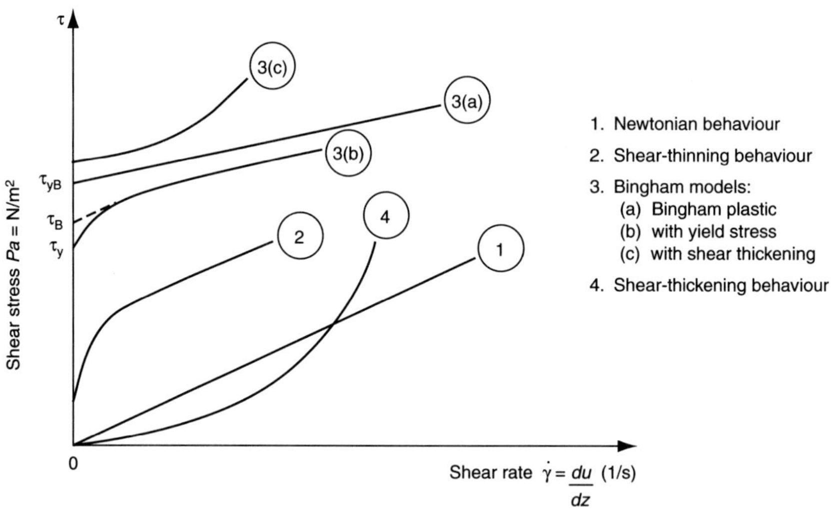


*Figure 5. Schematic representation of rheological models (after Verreet and Berlamont, 1988 – also see Table 7 for symbols). Reproduced by permission. From* Encyclopaedia of fluid mechanics, *© 1988, Gulf Publishing Company, Houston, Texas. All rights reserved*

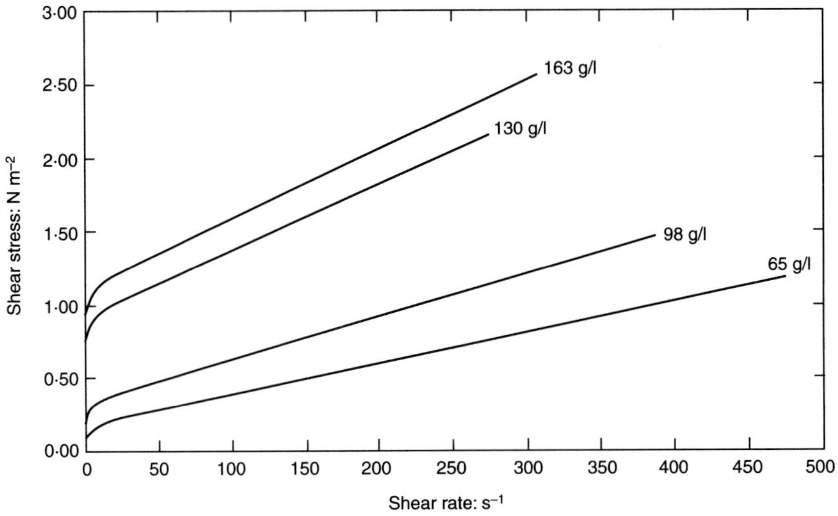


*Figure 6. Shear stress/shear rate flow curves for different concentrations*

volume fraction of the suspension. Deviation from Newtonian behaviour occurs at a dry mass concentration of around $10\,kg\,m^{-3}$ (or $10\,gl^{-1}$). Typical results for mud:water mixtures are shown in Figure 6. These curves relate to the 'up curve', i.e. conditions for the start-up of flow. In a shear-thinning material, the apparent viscosity decreases with increasing shear rate. A viscoelastic material has properties of liquids (dissipation of viscous energy by means of flow) and of solids (storage of elastic energy). Thixotropy, a continuous decrease in apparent viscosity with time, under constant shear and subsequent recovery when the shearing is stopped, is shown by a hysteresis loop for increasing or decreasing shear rate: there is no longer a unique relation between shear stress, $\tau$, and shear rate, $\dot{\gamma}$.

The non-Newtonian fluid properties of the mixtures are largely functions of the concentration and the type of clay in the fluid matrix. The apparent viscosity and Bingham yield stress depend on the shear rate and increase exponentially with sediment concentration (O'Brien and Julien, 1986; Jones and Golden, 1990). The yield stress and viscosity can increase by three orders of magnitude as the volume concentration of sediment changes from 10–40% (O'Brien and Julien, 1988). Recently, Van Kessel and Blom (1998) have studied the applicability of rheological models to a natural and an artificial mud. They concluded that the Bingham plastic model is only suitable for describing the measurements at shear rates in excess of $20\,s^{-1}$, because of the failure to adequately describe the changes in bed structure at lower shear rates. A similar conclusion can be drawn from Figure 6. However, the Bingham plastic model (Figure 5, model 3(a)) adequately represents the conceptual shear stress/shear rate relationship for most practical problems. Models 3(b) and 3(c) are variants on this model with, respectively, a yield stress and shear thickening properties (Verreet and Berlamont, 1988).

### *Procedure*

The following approach can be taken to characterise the rheological properties of the sediment.

*Field and laboratory measurements*

1. Use in-situ rheometers to measure the response of a cohesive sediment bed to shear.
2. Determine the shear stress/shear rate curve for a sample of material in a laboratory rheometer. This should be a controlled stress rheometer which does not force the material to move. Test at low shear rates ($<1/s$) to obtain representative results for most sediment transport applications (MAST G6M, 1993).

*No field or laboratory measurements*

3. Use the flow curve presented in Figure 6 to determine an approximate shear stress/shear rate response for a given concentration.

## 2.7. CONTAMINANTS

The aim of this section is to provide an introduction as to the role of contaminants in estuarine sediments. Further information can be found in Salomons *et al.* (1988).

### 2.7.1. *Chemical pollution*

There is much uncertainty about the role of contaminants on the dynamics of estuarine muds. It is known that contaminants are strongly associated with cohesive sediments (e.g. Grant and Gust, 1987). It is also known that biological processes and algal mats can have a significant role in altering expected physical properties of muds and their subsequent behaviour and transport. The potential impact of environmental concentrations of contaminants on key biota and ecological processes which both stabilise and destabilise sediments are not well understood. The biological effect of contaminants will depend on the exposure of biota (uptake from the sediments and surrounding water) and the actual toxicological effects. The dynamics of estuarine muds and contaminant-sediment interactions will play a role in determining the uptake and biological effects of contaminants.

Contaminants are substances in the environment that occur above expected background concentrations through their introduction by man or owing to geochemical extremes. A contaminant subsequently becomes a pollutant when it reaches concentrations that cause a biological effect. Some naturally occurring elements, such as several trace metals (eg. copper and zinc) are essential and 'beneficial' at certain concentrations. With increasing concentrations, metals may become inhibitory and ultimately toxic. Compounds, such as organic micro-contaminants, are manmade and at sufficiently high concentrations their presence can be deleterious.

Contaminants may enter estuarine systems from a variety of sources and by many routes with the relative contributions from different sources varying between contaminants and their patterns of use. Fluvial inputs from inland discharges and atmospheric deposition are often a dominant input of contaminants to estuaries. Inputs may also arise, however, from activities within the estuary such as shipping. Therefore, a wide suite of contaminants may be expected to be present in estuarine muds from both historical and contemporary discharges. Contaminants generally recognised to be of concern in estuarine environments include metals (such as mercury, lead, copper, cadmium and zinc), radioactive waste, organo-metal derivatives (e.g. tributyltin), organochlorinated pesticides, organophosphorus insecticides, polychlorinated biphenyls, herbicides, oil and detergent products, polyaromatic hydrocarbons, dioxins and furans. The extent of contamination within estuaries will vary greatly.

Estuarine muds are ionic in nature as mentioned previously. The ionic nature of cohesive sediments makes it susceptible for adsorption of contaminants from the surrounding water. The contaminants may be associated with the clay mineral grains with small grain sizes giving large surface areas for adsorption (on the surface) or contaminants may be absorbed (within) into the clay minerals. Contaminants are also associated with organic matter in sediments.

Cohesive sediments may act as a source or sink for toxic contaminants over time depending on the surrounding physico-chemical conditions. The importance of the physico-chemical conditions, such as sediment type, organic matter content, pH, salinity, redox and temperature to contaminant availability are well documented (Beeftink *et al.*, 1982; Kersten, 1988). High pH and positive redox values permit the formation of iron and manganese oxides and hydroxides, bring about coprecipitation and adsorption of metals onto mud surfaces, and generally reduce contaminant uptake. The lower the pH value, the greater the bioavailability of metals, such as mercury. Under reducing conditions (negative redox values), anaerobic sediments prevail and the presence of free sulphide at different concentrations will affect the pore water concentrations of metals and, therefore, their bioavailability. Organic chelators, inorganic ligands and other contaminant binding components of sediments may also have a dramatic influence on contaminant bioavailability through the effects of sorption and co-precipitation. Humic acids can also play a significant role in metal speciation and toxicity in estuarine systems. Complexation by organic ligands generally reduces toxicity. As a result toxicity is usually found to correlate best with free metal ion concentrations, which may be many orders of magnitude less than the total metal concentrations, and with the dissolved organic micro-contaminant concentrations.

Contaminated sediments pose a threat to biota as contaminants may be released from disturbed sediments through, for example, dredging operations, resuspension, transport and erosion processes or pore water escape during consolidation processes.

The sediment-contaminant interactions are in a dynamic equilibrium with the surrounding environment. Consequently, the exposure, uptake and subsequent contaminant burden of an organism depends on the contaminant concentration and speciation in muds and surrounding waters, the prevailing hydraulic regime coupled with biological factors such as the physiology and biochemistry of the organism, life history and the trophic position in the food web. This complex interaction of varying parameters means that a general assessment of the bioavailability of sediment-bound contaminants within estuarine environments is difficult to make. However, it is likely that contaminants play a role in the dynamics of sediments by altering mud properties.

### 2.7.2. *Biohazards*

Marine sediments provide a rich substrate for marine life in its abundant forms, both within, and on, the surface of the sediment. Cohesive sediment can contain high proportions of biomass and is thus an important part of the food web of sediment ecology. Hazardous bacteria and diseases may be present in organic rich sediments, particularly where anthropogenic (man-made) waste products are discharged or have accumulated, such as sewage and rubbish.

# Hydrodynamics

# 3. Hydrodynamics

## 3.1. GENERAL

The movement of mud in an estuary depends on the hydrodynamic forcing caused by currents, waves and pressure gradients, and on the gravitational forces on a sloping bed. Well-mixed estuaries have a relatively simple behaviour, and can be treated in a similar manner to coastal waters. The circulation in a stratified or salt-wedge estuary can be more complicated, with the flow at the surface sometimes being unrelated to that at the bed, and even in the opposite direction. The Coriolis effect can give rise to cross-estuary components of flow and water surface slopes in wide estuaries. The circulation patterns in estuaries are described in detail by Dyer (1998b).

Estuaries are dominated by tidal processes. The tidal wave progresses up the estuary and is partially reflected, resulting in an almost standing wave. In a pure standing wave the times of high and low water correspond to slack water, and the times of maximum flood and ebb current occur at mid-tide. In practice, this is only an approximation to the behaviour in real estuaries, as there is usually a phase difference between high/low water and slack water, and there is often a strong asymmetry between the flood and ebb tidal currents. Estuaries of a certain shape and size can act as a resonator and amplify the tidal rise and fall and the currents, as in the case of the Severn Estuary. The relative influence of the fresh water input from rivers and the tidal flushing varies markedly between estuaries, and also along the length of an individual estuary.

Waves generally have a smaller influence on the sediments in an estuary than on an open coast. This is particularly true if there is a shallow bar at the mouth of the estuary, which excludes most of the wave energy from the adjacent sea area. In this case, it is primarily waves generated locally within the estuary which exert an influence. The height and period of the waves will depend strongly on the length of water (the *fetch*) across which the wind is blowing, so that an along-estuary wind will create a greater wave effect than an across-estuary wind. Waves in estuaries will tend to be small in height and period, so that their influence is restricted to the shallow mud-flat zone in depths less than about 1 m. Nevertheless, within this zone their effect can be very significant, and the wave-dominated zone will be advected across the mud-flats as the tide rises and falls.

The salinity (and hence water density) in an estuary may have strong horizontal gradients; in cases with an important river inflow the salinity may vary from 0 to 35 ppt along the length of the estuary. These give rise to pressure gradients, in addition to the pressure gradients caused by tidal water surface slopes, which can drive flows.

Estuaries often feature channels with steep sides, and fluid mud or an unconsolidated deposit may flow down these slopes under the effect of gravity. Even on more gently sloping mudflats a low-density deposit can flow downhill.

The main hydrodynamic parameter that controls the erosion, suspension and deposition of muds is the *bed shear-stress*, which is the frictional force exerted by the flow per unit area of bed. Methods of calculating the bed shear-stress produced by currents and waves, separately or in combination, are therefore needed. The methods are broadly similar to those used for beds of coarser sediments, but with the difference that the flow is usually assumed to be hydrodynamically smooth for muds whereas it is usually treated as hydrodynamically rough for sands and gravels. The formulae described below are adaptations to the case of hydrodynamically smooth flow of those presented in Soulsby (1997).

Note, however, that in certain cases a mud bed may be very rough as a result of trawling activities or bioturbation; for example, box cores of muddy sediments taken at depths of up to 60 m in the Clyde Sea, western Scotland, were found to have had a very disturbed irregular surface with lumps of firm mud several centimetres in height. Bed roughnesses in estuaries may also sometimes be large, for example in areas with a network of fine drainage channels. In these cases the *total* bed shear-stress will be governed by the large roughness which incorporates the form-drag created

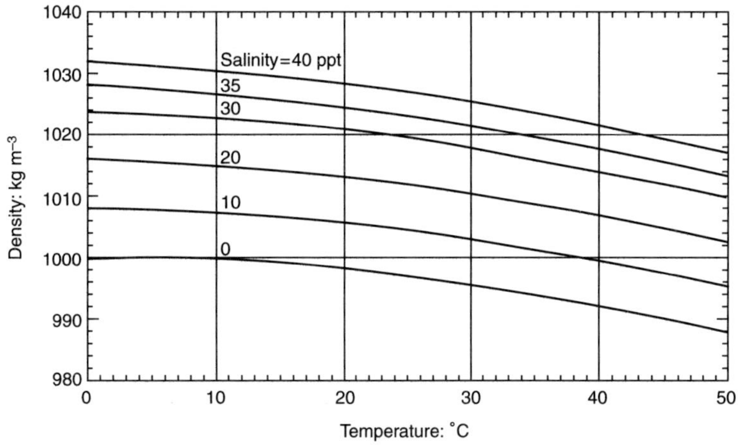


*Figure 7. Density of water (Soulsby, 1997)*

by the pressure field around the roughness elements. Nevertheless, the erosion and deposition of the mud will depend more on the smooth flow over the intervening areas, and the smooth-hydrodynamic equations given below can be used.

Many of the methods presented in this section and elsewhere require values for the water density $\rho$ and the kinematic viscosity $\nu$. Both of these depend on the temperature and salinity of the water. Their values can be obtained from the curves given in Figures 7 and 8 (Soulsby, 1997) or from the SandCalc software package (Section 1.2).

The presence of sediment in the water will increase both the bulk density and the bulk viscosity of the suspension. The bulk density of the suspension is given by

$$\rho_B = \rho + C_M\left(\frac{\rho_s - \rho}{\rho_s}\right) \tag{3.1}$$

where $\rho_B$ = bulk density of suspension
$\rho$ = density of water, excluding sediment
$\rho_s$ = density of sediment grains
$C_M$ = mass concentration (dry density) of sediment (mass/volume)

Conversions to and from other measures of the concentration are given in Table 2 and Figure 1.

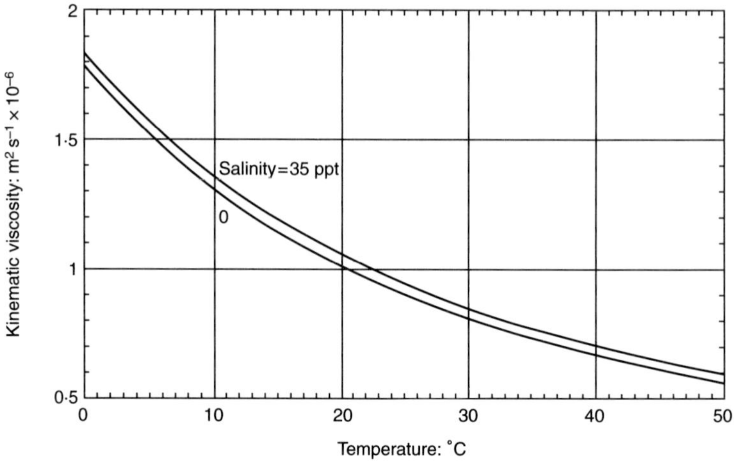


*Figure 8. Kinematic viscosity of water (Soulsby, 1997)*

A theoretical derivation of the effect of suspended sediment on the bulk viscosity, accurate to second order in volume concentration, was derived by Batchelor and Green (1972). Written in terms of the quantities defined above, this expression is:

$$\nu_B \rho_B = \nu\rho\left[1 + 2{\cdot}5\left(\frac{C_M}{\rho_s}\right) + 7{\cdot}6\left(\frac{C_M}{\rho_s}\right)^2\right] \qquad \text{SC (3.2)}$$

where $\nu_B$ = bulk kinematic viscosity of suspension
$\nu$ = kinematic viscosity of water, excluding sediment

Equation (3.2) is theoretically valid up to a volume concentration $C$ of sediment of about 10%, that is, a mass concentration or dry density $C_M$ of about 265 kg m$^{-3}$. However, in practice the flow becomes non-Newtonian for natural mud suspensions at mass concentrations in excess of about 10 kg m$^{-3}$, and a full rheological treatment should be used (see Section 2.6.2).

## 3.2. CURRENTS

### Knowledge

A current flowing in an estuary experiences friction with the bed, which forms a turbulent *boundary layer*. Within this the current speed increases with height from zero at the sea bed to a maximum at or near the water surface. The variation of current speed with height above the bed is known as the current *velocity profile*.

Within the bottom few metres the current speed $U$ varies with height $z$ above the bed according to the logarithmic velocity profile

$$U(z) = \frac{u_*}{\kappa} \ln\left(\frac{z}{z_0}\right) \qquad \text{SC (3.3)}$$

where $u_*$ = friction velocity
$z_0$ = bed roughness length
$\kappa$ = Von Karman's constant = 0·40

The bed shear-stress $\tau_0$ can be obtained from the friction velocity $u_*$ through the relationship

$$\tau_0 = \rho u_*^2 \qquad \text{SC (3.4)}$$

where $\rho$ = density of water

For *hydrodynamically smooth* flow (see below for definition), as is found typically in muddy estuaries, the bed roughness length depends only on $u_*$ and the kinematic viscosity $\nu$ of the water, according to the relationship

$$z_0 = \nu/(9u_*) \qquad (3.5)$$

Substituting Equation (3.5) into (3.3) and re-arranging gives the logarithmic profile in a form commonly quoted for smooth flows:

$$\frac{U(z)}{u_*} = 5{\cdot}5 + 2{\cdot}5 \ln\left(\frac{u_* z}{\nu}\right) \qquad \text{SC (3.6)}$$

If there are significant roughness elements on the bed then the flow may be hydrodynamically rough or transitional, and a more complicated expression for $z_0$ must be used, such as that of Christofferson and Jonsson (1985):

$$z_0 = \frac{k_s}{30}\left[1 - \exp\left(\frac{-u_* k_s}{27\nu}\right)\right] + \frac{\nu}{9u_*} \qquad \text{SC (3.7)}$$

Here $k_s$ is the *Nikuradse roughness*, given by

$$k_s = 2{\cdot}5d_{50} \tag{3.8}$$

if the roughness is due to coarse grains of median diameter $d_{50}$ (Soulsby (1997) also gives some alternative expressions). For other kinds of roughness, such as worm casts or fine drainage channels, $k_s$ can be set to the height of protrusion as an approximation.

Equation (3.6), and also (3.3) with (3.7), are implicit equations – they cannot be solved directly for $u_*$ in terms of $U$. Thus if $U(z)$ is known and $u_*$ is to be calculated, an iterative computational scheme is required.

In many cases the *depth-averaged current speed* $\bar{U}$ is required or supplied. This is related to the velocity profile $U(z)$ through the definition

$$\bar{U} = \frac{1}{h}\int_0^h U(z)\,dz \tag{3.9}$$

where $\bar{U}$ = depth-averaged current speed
$h$ = water depth
$U(z)$ = current speed at height $z$
$z$ = height above sea bed

The lower limit of the integral has to be changed from 0 to $z_0$ if the velocity profile tends to zero at a height $z = z_0$ (e.g. for a logarithmic velocity profile).

Substituting Equation (3.3) for $U(z)$ into Equation (3.9) gives the following relationship between the depth-averaged current speed and the friction velocity

$$\bar{U} = \frac{u_*}{\kappa}\left[\ln\left(\frac{h}{z_0}\right) - 1\right] \qquad \text{SC}\,(3.10)$$

This is based on the assumption that Equation (3.3) holds throughout the water depth, which is not usually the case, but nevertheless Equation (3.10) is widely used.

Combining Equations (3.6) and (3.10) allows $u_*$ to be calculated iteratively, if $\bar{U}$ is known, for smooth flows. Equations (3.6) and (3.10) have been used to construct the curves in

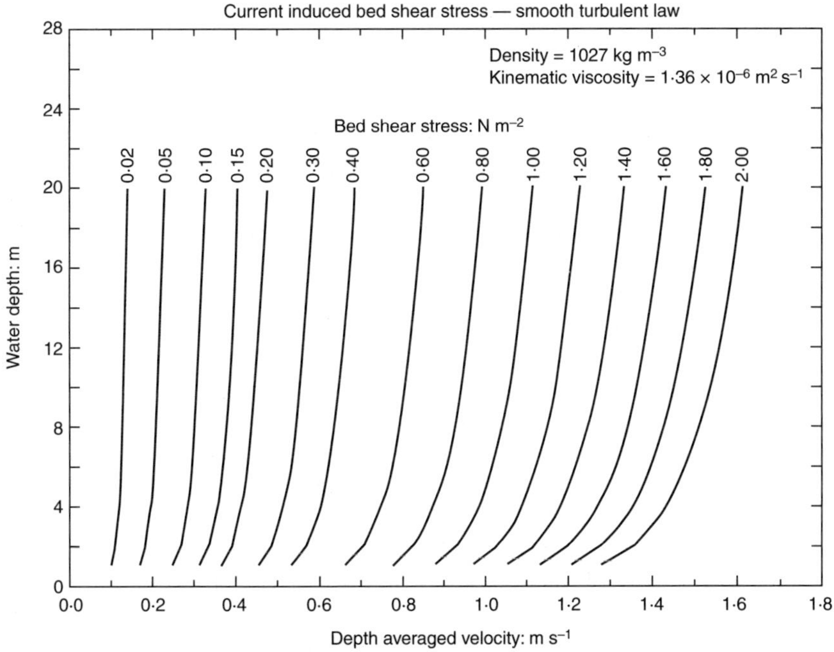


*Figure 9. Current induced shear-stress (smooth turbulent law) as a function of depth-averaged flow velocity and water depth*

Figure 9, which allow the bed shear-stress $\tau_0$ to be calculated from the depth averaged velocity $\bar{U}$ for a given water depth, and for a water density of $\rho = 1027\,\mathrm{kg\,m^{-3}}$ and kinematic viscosity $\nu = 1{\cdot}36 \times 10^{-6}\,\mathrm{m^2\,s^{-1}}$. Likewise, for transitional flows (and also for smooth and rough flows) a combination of Equations (3.3), (3.7) and (3.10) allows $u_*$ to be calculated iteratively from $\bar{U}$.

The decision as to whether the flow is smooth, transitional or rough depends on the grain Reynolds number $u_* k_s/\nu$:

| | |
|---|---|
| $u_* k_s/\nu \leqslant 5$ | smooth |
| $5 < u_* k_s/\nu < 70$ | transitional |
| $u_* k_s/\nu \geqslant 70$ | rough |

If the bed appears to be smooth (as will be the case in many mud problems), then it is best to use Equation (3.6), because it is then not necessary to estimate a value for $k_s$. In this case, $u_*$ can be calculated from $\bar{U}$ using Figure 9.

In all cases, the following points should be considered:

- if there is a strong vertical salinity gradient (e.g. in a salt-wedge estuary), then Equation (3.3) applies only close to the bed, and the integration for $\bar{U}$ given by Equation (3.10) is invalid
- if the suspended mud itself causes a strong vertical density gradient, then results from all the above methods may be inaccurate
- at times within one hour either side of slack water, acceleration effects are important and Equation (3.3) is inaccurate
- a fluid mud layer at the bed will strongly modify the near-bottom velocities and Equations (3.3) and (3.6) are invalid
- concentrations of mud in excess of about $50\,\text{kg}\,\text{m}^{-3}$ (dry density) will increase the kinematic viscosity by more than 5%, and corrections (possibly non-Newtonian) should be made (see Section 2.6.2, Rheology)
- the logarithmic velocity profile applies to straight uniform flow over an 'infinite' flat bed, and hence can be applied validly on an expanse of mudflat, but topographic effects may cause large departures in the vicinity of natural or dredged channels, near to manmade or natural structures, or on sharp bends
- field values of $z_0$ are given by Soulsby (1997) as 0·2 mm (mud), 0·7 mm (mud/sand) and 0·05 mm (silt/sand).

## *Procedure*

The bed shear-stress due to currents can be calculated as follows:

1. Obtain the temperature and salinity of the water, and use Figures 7 and 8 to calculate the density $\rho$ and kinematic viscosity $\nu$ of the water.

2. Measure the current speed $U(z)$ at 3 to 6 heights above the bed within the bottom 1 m of the water column, taking an average speed over about 10 minutes, at about 30 minute intervals through a tidal cycle. For each profile, plot the speed $U(z)$ versus the natural logarithm of height $\ln(z)$. According to Equation (3.3), the data should fall approximately on a straight line. Perform a linear regression of the form $y = mx + c$, with $U(z)$

as $y$, and $\ln(z)$ as $x$. Making use of Equation (3.3), calculate

$$u_* = 0{\cdot}40\,m \tag{3.11}$$

$$z_0 = \exp(-c/m) \tag{3.12}$$

Check that the correlation coefficient obtained from the linear regression is close to 1, as a value in excess of 0·99 is needed for an accurate estimate of $u_*$. A methodology for estimating the statistical errors is given by Wilkinson (1984).

Calculate bed shear-stress $\tau_0$ from Equation (3.4).

This method can be used over smooth, transitional or rough beds, and gives the *total* bed shear-stress. The value of $z_0$ should be approximately $\nu/(9u_*)$ if the flow is smooth, larger if it is transitional or rough.

3. If only a measurement of current speed at one height is available, and the bed is believed to be smooth, then Equation (3.6) can be solved iteratively to obtain $u_*$.

   If the bed is rough, then estimate a value of the height $k_s$ of the roughness elements, and solve Equation (3.3) iteratively with Equation (3.7).

   In either case, use Equation (3.4) to calculate $\tau_0$ from $u_*$.

4. If only a depth-averaged current speed $\bar{U}$ is available, and none of the points listed at the end of the *Knowledge* section is significant, obtain $\tau_0$ from Figure 9, which is based on Equation 3.6.

5. Alternatively, use SandCalc version 2.00 under Hydrodynamics–Currents–Skin Friction–Logarithmic Smooth (if bed is known to be smooth) or 'Smooth or Rough' if coarser grains are present. A grain diameter of $d_{50} = 0{\cdot}01$ mm can be input if the bed is only known to be a smooth mud; this will not significantly affect the results since the flow will be hydrodynamically smooth.

*Example 3.1. Bed shear-stress for a current*

To calculate the bed shear-stress produced by a current if the depth-averaged velocity is known, for water at 10°C and 35 ppt, obtain estimates of the following parameters:

| | | |
|---|---|---|
| depth-averaged current speed | $\bar{U}$ | $0{\cdot}5\,m\,s^{-1}$ |

| | | |
|---|---|---|
| water depth | $h$ | 5 m |
| grain size | $d_{50}$ | 0·03 mm |

Calculate water density and kinematic viscosity from Figures 7 and 8, or from SandCalc.

| | | |
|---|---|---|
| water density | $\rho$ | $1027\,\mathrm{kg\,m^{-3}}$ |
| kinematic viscosity | $\nu$ | $1{\cdot}36 \times 10^{-6}\,\mathrm{m^2\,s^{-1}}$ |

First, assume hydrodynamically smooth flow. Use Equations (3.9) and (3.6) and iterate for $u_*$, or use SandCalc–Currents–Skin Friction–Logarithmic Smooth.

| | | |
|---|---|---|
| friction velocity | $u_*$ | $0{\cdot}0164\,\mathrm{m\,s^{-1}}$ |
| bed shear-stress | $\tau_0$ | $0{\cdot}276\,\mathrm{N\,m^{-2}}$ |

Alternatively, take account of the roughness of the silt bed. Solve Equations (3.3) iteratively with Equation (3.7) to obtain $u_*$, or use SandCalc–Currents–Skin Friction or Rough.

| | | |
|---|---|---|
| friction velocity | $u_*$ | $0{\cdot}0164\,\mathrm{m\,s^{-1}}$ |
| bed shear-stress | $\tau_0$ | $0{\cdot}276\,\mathrm{N\,m^{-2}}$ |

In this example the assumption that the flow was smooth was correct, and the presence of the silt grains does not significantly affect the answers.

## 3.3. WAVES

### *Knowledge*

Waves in sufficiently shallow water produce an oscillatory velocity at the sea bed, which acts on the sediments. 'Sufficiently shallow' in this context means approximately

$$h < 0{\cdot}1gT^2 \tag{3.13}$$

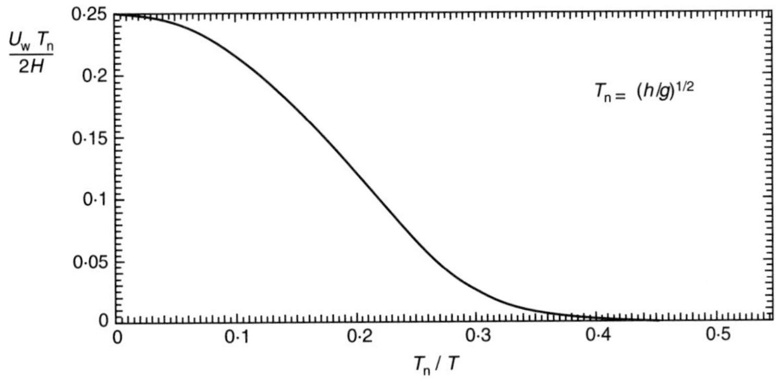


*Figure 10. Bottom velocity for monochromatic waves (Soulsby and Smallman, 1986)*

or, alternatively

$$h < 10H_s \tag{3.14}$$

where $h$ = water depth
$H_s$ = significant wave height
$T$ = wave period
g = acceleration due to gravity

The amplitude $U_w$ of the wave orbital velocity just above the bed due to a monochromatic (single frequency) wave of height $H$ and period $T$ in water of depth $h$ is

$$U_w = \frac{\pi H}{T \sinh(kh)} \qquad \text{SC (3.15)}$$

where sinh is the hyperbolic sine, $k = 2\pi/L$ is the wave number, and $L$ is the wavelength. Figure 10 gives a curve marked 'Monochromatic' based on Equation (3.15) from which $U_w$ can be calculated directly from the input parameters $H$, $T$, $h$ and g, via the quantity $T_n = (h/g)^{1/2}$ (Soulsby and Smallman, 1986).

In estuaries and the sea, a spectrum of waves of different heights, periods and directions will be present. The waves are usually characterised by their significant height $H_s$ and zero-crossing period $T_z$ (or, almost equivalent, $T_m$). For the purpose of calculating bed shear-stresses to determine mud behaviour, a representative monochromatic wave must be chosen. A good choice for this is the wave of height $H_{rms}$ and period $T_p$, where $H_{rms}$ is the root-mean-square wave height obtained from $H_s$ by the relationship

$$H_{rms} = H_s/\sqrt{2} \qquad \text{SC (3.16)}$$

and $T_p$ is the peak period of the spectrum obtained from $T_z$ by the relationship

$$T_p = 1{\cdot}28 T_z \qquad \text{SC (3.17)}$$

The monochromatic wave of height $H = H_{rms}$ and period $T = T_p$ matches both the energy density of the spectral waves and the period at which the energy is greatest.

As with currents, the most important hydrodynamic property of waves for sediment transport purposes is the bed shear-stress they produce. This is oscillatory in the case of waves, having an amplitude $\tau_w$. It is usually obtained from the bottom orbital velocity $U_w$ of the waves via the wave friction factor $f_w$, defined by

$$\tau_w = \frac{1}{2}\rho f_w U_w^2 \qquad \text{SC (3.18)}$$

The wave friction factor is dependent on whether the flow is laminar, smooth turbulent, or rough turbulent, which in turn depends on the wave Reynolds number $R_w$ and the relative roughness $r$:

$$R_w = \frac{U_w A}{\nu} \qquad \text{SC (3.19)}$$

$$r = \frac{A}{k_s} \qquad \text{SC (3.20)}$$

where $U_w$ = bottom orbital velocity amplitude
$A = U_w T/2\pi$ = semi-orbital excursion
$T$ = wave period
$\nu$ = kinematic viscosity
$k_s$ = Nikuradse equivalent sand grain roughness (see Equation 3.8)

The smooth bed friction factor $f_{ws}$ can be calculated from

$$f_{ws} = BR_w^{-N} \qquad \text{SC (3.21)}$$

where

$$B = 2, \qquad N = 0{\cdot}5$$
$$\text{for } R_w \leqslant 5 \times 10^5 \text{ (laminar)} \qquad \text{SC (3.22)}$$

$$B = 0{\cdot}0521, \quad N = 0{\cdot}187$$
$$\text{for } R_w > 5 \times 10^5 \text{ (smooth turbulent)} \qquad \text{SC (3.23)}$$

The rough bed friction factor $f_{wr}$ can be calculated from the expression given by Soulsby (1997):

$$f_{wr} = 0{\cdot}237r^{-0{\cdot}52} \qquad \text{SC (3.24)}$$

Alternative formulae and further details are discussed in Soulsby (1997). The selection of laminar, smooth or rough friction factor is described below under *Procedure*.

The most intense disturbance of sediment is often found beneath breaking waves. Most existing theoretical studies, including the methods presented above, apply to non-breaking waves because the treatment of the hydrodynamics is simpler. It is common practice to apply the results from non-breaking waves to the case of breaking and broken waves, although this is not strictly justified. Some support for this approach is provided by the experiments of Deigaard *et al.* (1991), who showed that the bed shear-stresses in the surf zone were not *on average* very different from those of unbroken offshore waves, but they exhibited much greater wave-to-wave variability, so that occasional very large values could occur. Also, the neglect of wave breaking could lead to an underestimate of suspended sediment concentrations in the upper half of the water column in the surf zone, where turbulence derived from the breaking process is important. The role of breaking waves on sediment transport processes in general still requires further research.

## *Procedure*

The bed shear-stress due to waves can be calculated as follows:

1. Obtain values of water depth $h$ and significant wave height $H_s$ and zero-crossing period $T_z$, or similar measures of wave properties.
2. Convert to an equivalent monochromatic wave of height $H = H_{rms}$ and period $T = T_p$ using Equations (3.16) and (3.17).
3. Calculate the bottom orbital velocity amplitude $U_w$ of the waves using Figure 10, or by solution of Equation (3.15). The latter requires an iterative solution of the wave dispersion relation to find the wavenumber $k$, see Section 4.3 of Soulsby (1997).
4. Obtain the temperature and salinity of the water, and use Figures 7 and 8 to calculate the density $\rho$ and kinematic viscosity $\nu$ of the water.
5. If the bottom is smooth and featureless, as is frequently the case for freshly deposited muds, calculate the wave Reynolds number $R_w$ from Equation (3.19). Then calculate the smooth bed friction factor $f_{ws}$ using Equations (3.21) and (3.22) if $R_w \leqslant 5 \times 10^5$, or Equations (3.21) and (3.23) if $R_w > 5 \times 10^5$.
6. Calculate the amplitude $\tau_w$ of the bed shear-stress using Equation (3.18) with $f_w = f_{ws}$.
7. If the bottom is rough due to the presence of coarse grains, worm casts, drainage channels, etc., estimate the value of the Nikuradse roughness $k_s$ as $2{\cdot}5\, d_{50}$ for grains, or as height of protuberance of other roughness elements.
8. Calculate the relative roughness $r$ using Equation (3.20), and the rough bed friction factor $f_{wr}$ from Equation (3.24). If $f_{wr}$ is larger than $f_{ws}$ found in step 5 above, then take $f_w = f_{wr}$ and calculate $\tau_w$ using Equation (3.18). This gives the total bed shear-stress, including form drag on the roughness elements.
9. Even if the bed is rough (i.e. if $f_{wr} > f_{ws}$), it is probably better to use the smooth-bottom bed shear-stress given in step 6 above to investigate threshold of motion, erosion rates and deposition rates (see Sections 4.2, 4.3 and 8).
10. If the waves are breaking or broken, use the method given above, but treat the results with more caution and be aware that occasional very large values of bed shear-stress will occur.
11. Alternatively, use SandCalc version 2.00 under Hydrodynamics–Waves–Orbital Velocity–Monochromatic to obtain $U_w$, and Hydrodynamics–Waves–Skin Friction–Soulsby to obtain $\tau_w$.

*Example 3.2. Bed shear-stress for waves*
To calculate the bed shear-stress produced by waves, in the same water conditions as Example 3.1 (10°C, 35 ppt, $\rho = 1027\,\mathrm{kg\,m^{-3}}$, $\nu = 1{\cdot}36 \times 10^{-6}\,\mathrm{m^2\,s^{-1}}$), obtain estimates of the following parameters:

| | | |
|---|---|---|
| water depth | $h$ | 5 m |
| significant wave height | $H_s$ | 1 m |
| zero-crossing period | $T_z$ | 3·5 s |
| grain-size | $d_{50}$ | 0·03 mm |

Convert to an equivalent monochromatic wave using Equations (3.16) and (3.17), or SandCalc.

| | | |
|---|---|---|
| RMS wave height | $H_{rms}$ | 0·707 m |
| peak period | $T_p$ | 4·5 s |

Calculate the bottom orbital velocity amplitude using Figure 10 or SandCalc–Waves–Orbital Velocity–Monochromatic.

| | | |
|---|---|---|
| orbital velocity amplitude | $U_w$ | $0{\cdot}328\,\mathrm{m\,s^{-1}}$ |

Follow Steps 5 to 9 of *Procedure* above, or use SandCalc–Hydrodynamics–Waves–Skin Friction, to obtain bed shear-stress.

| | | |
|---|---|---|
| bed shear-stress amplitude | $\tau_w$ | $0{\cdot}465\,\mathrm{N\,m^{-2}}$ |

In this case the Reynolds number $R_w = 5{\cdot}7 \times 10^4$ and the flow is laminar.

## 3.4. COMBINED WAVES AND CURRENTS

### *Knowledge*

In many cases both currents and waves make significant contributions to the bed shear-stress. The resulting bed shear-stress consists of a steady component due to the current together with an oscillatory component due to the waves. If the current and wave velocities over a smooth bed are sufficiently small that the flow remains laminar, then the combined bed shear-stresses are simply a linear addition of the laminar current-alone and wave-alone shear-stresses. However, in stronger currents and waves the flow will be turbulent, and the turbulence generated in the current and wave boundary layers then adds in a non-linear fashion. Because of this, the mean and oscillatory components of the stress are enhanced beyond the values which would result from a simple linear addition of the wave-alone and current-alone stresses, as shown schematically in Figure 11.

The bed shear-stress $\tau(t)$ varies through a wave cycle in the manner shown schematically by the dotted locus in Figure 11.

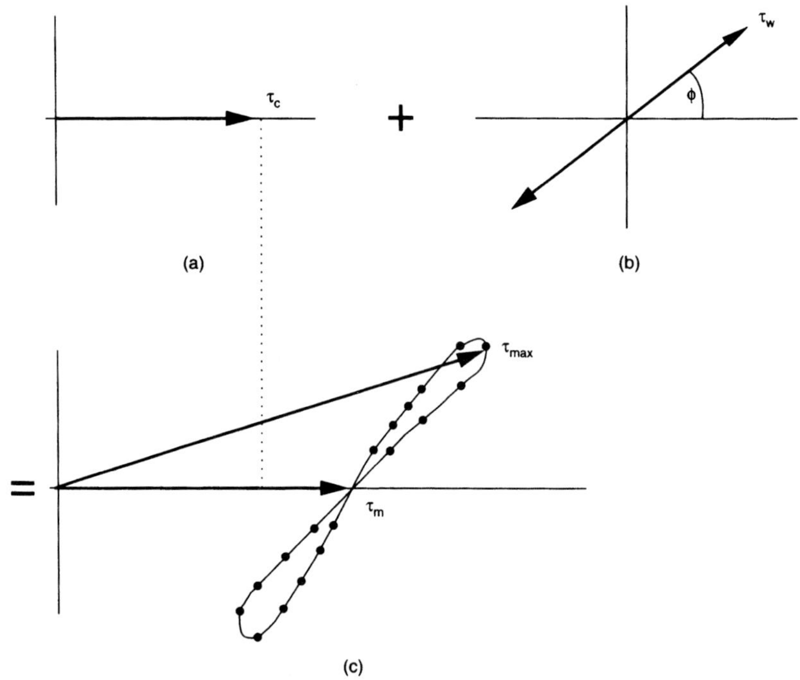


*Figure 11. Schematic representation of non-linear interaction of wave and current bed shear-stresses. Reproduced from Soulsby* et al. *(1993), Wave-current interaction within and outside the bottom boundary layer,* Coastal Engng, ***21****, 41–69, with permission from Elsevier Science*

The most important quantities for use in sediment calculations are:

- the mean value, $\tau_m$, over the wave-cycle
- the maximum value, $\tau_{max}$, during the wave-cycle
- the root-mean-square value, $\tau_{rms}$, taken over the wave-cycle.

The mean, $\tau_m$, is used for determining the friction governing the current, and for determining diffusion of sediment into the outer flow; the maximum, $\tau_{max}$, is used to determine the threshold of sediment motion, and diffusion very near the bed; and $\tau_{rms}$ is a good average measure of the shear-stress, particularly useful in random waves.

More than twenty different theories and models have been proposed to describe this process, but almost all of them have concentrated on the case of a rough turbulent flow, as would be found over a coarse sand or gravel bed, rather than the smooth turbulent flow commonly found over a mud bed. Some of these theories were discussed and inter-compared by Soulsby *et al.* (1993). A parametric fitting method for some of the theories is presented in Soulsby (1997).

The theories differ markedly in their predictions, and in the goodness of their fit to data (Soulsby *et al.*, 1993). The best performing analytical theories were those of Fredsøe (1984) and Grant and Madsen (1979), although both are complicated to use. In view of this, the following equation, which has the same general form as that fitted to the theories, was proposed by Soulsby (1995) as a direct fit to 61 laboratory measurements and 70 field measurements of the cycle-mean bed shear-stress $\tau_m$ (all for rough beds):

$$\tau_m = \tau_c\left[1 + 1{\cdot}2\left(\frac{\tau_w}{\tau_c + \tau_w}\right)^{3\cdot2}\right] \qquad \text{SC (3.25)}$$

in which $\tau_c$ and $\tau_w$ are the bed shear-stresses which would occur due to the current alone and to the wave alone, respectively.

The corresponding expression for $\tau_{max}$ is given by a vector addition of $\tau_m$ from Equation (3.25) and $\tau_w$ obtained through Equations (3.21) to (3.24):

$$\tau_{max} = [(\tau_m + \tau_w\cos\phi)^2 + (\tau_w\sin\phi)^2]^{1/2} \qquad \text{SC (3.26)}$$

where $\phi$ = angle between current direction and direction of wave travel.

This is based on an assumption that the enhancement of the oscillatory component of stress caused by the current-induced turbulence is negligible (but note that many of the more sophisticated theories do account for this). Similarly, the root-mean-square bed shear-stress is given by:

$$\tau_{rms} = (\tau_m^2 + \tfrac{1}{2}\tau_w^2)^{1/2} \tag{3.27}$$

As indicated, the theories and Equation (3.25) are all designed for rough-bed conditions. A comprehensive set of laboratory measurements made by Arnskov *et al.* (1993) of bed shear-stress generated by combined waves and currents over a smooth solid bed gives some indication of the flow behaviour over smooth mud beds (Figure 12). They found a significant non-linear enhancement of $\tau_m$ for wave-dominated conditions, but no enhancement for current-dominated conditions. Their measurements showed little or no enhancement of $\tau_{max}$, contrary to the predictions of most rough-bed theoretical models. They also found evidence that current-generated turbulence was suppressed by large waves. They tested the rough-bed theoretical model of Fredsøe (1984) against this set of smooth-bed data and found it gave poor agreement. Equation (3.25) also gives a poor fit to this data.

Although the general formulation used to derive Equation (3.25) was based on data for rough beds, it seems reasonable to extend the basic non-linearity it expresses to the case of smooth beds by fitting its two free coefficients to the data of Arnskov *et al.* (1993). Soulsby (2000) thus obtained the following equation for smooth beds:

$$\tau_m = \tau_c\left[1 + 9\left(\frac{\tau_w}{\tau_c + \tau_w}\right)^9\right] \tag{3.28}$$

The values of $\tau_c$ and $\tau_w$ are calculated using the smooth-turbulent methods given in Sections 3.2 and 3.3. The calculations of $\tau_{max}$ and $\tau_{rms}$ are again given by Equations (3.26) and (3.27).

Figure 12 shows that Equations (3.28) and (3.26) give a reasonably good fit to the smooth-bed measurements of $\tau_m$ and $\tau_{max}$ made by Arnskov *et al.* (1993) and Sleath (1990) for waves

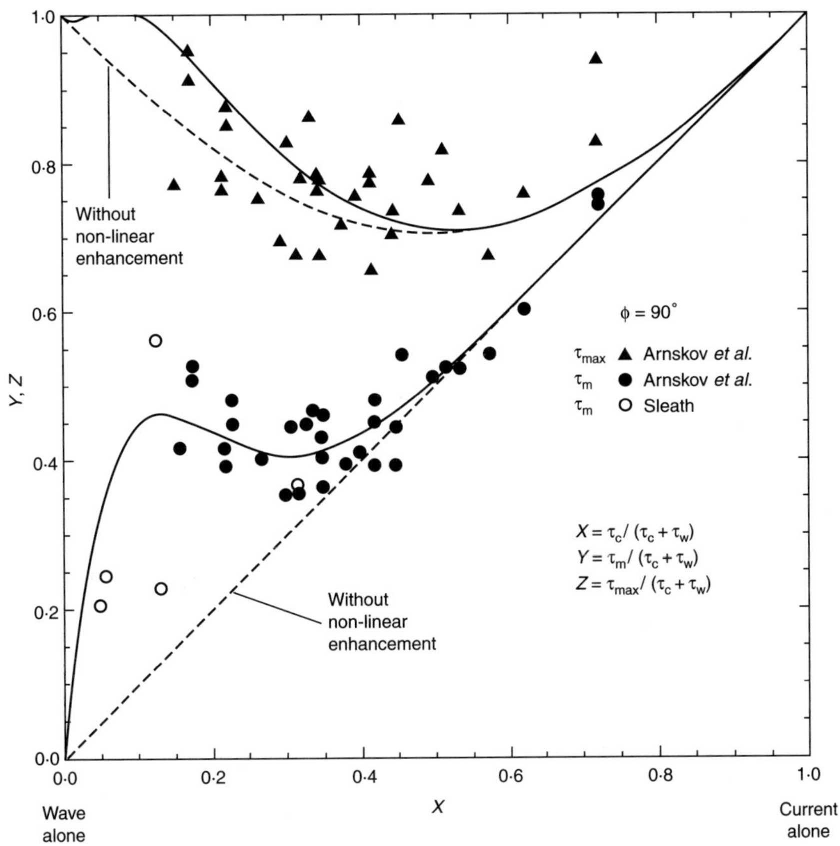


*Figure 12. Comparison of fitted curves with data for the mean $\tau_m$ and maximum $\tau_{max}$ values of the wave current shear-stress; waves and currents crossing at right angles (Soulsby, 2000). Data from Arnskov* et al. *(1993) and Sleath (1990)*

travelling at 90° to a current. Similar agreement was found for angles of 72° and 108°. The more sophisticated existing theories are unlikely to give better overall agreement, so Equations (3.28) with (3.26) and (3.27) are recommended for present purposes.

Probably the most suitable theoretical model for wave-plus-current bed shear-stresses over mud beds is that of Myrhaug and Slaattelid (1990), which is valid for rough, smooth and transitional turbulent flows. The formulation is too complicated to present here, so the reader is referred to the original paper.

Professor Myrhaug has developed an iterative computational solution of the equations presented. This formulation has a stronger theoretical basis than the empirical approach given by Equation (3.28).

An additional factor which is sometimes considered in making estimates of bed shear-stress with mud in suspension is the relative influence of the suspended sediment on the bulk viscosity (Equation 3.2) versus the influence of the suspension (or suspension gradient) in producing drag reduction (e.g. Li and Gust, 2000). However, Wang *et al.* (1998) concluded that over a smooth bed the effect of damping turbulence by a clay suspension was counterbalanced by the effect of increasing viscous resistance; therefore no net drag reduction occurred and the effects can be ignored.

## *Procedure*

The bed shear-stress under combined wave-current conditions can be calculated as follows.

1. Use the *Procedure* in Section 3.2 to calculate the current-alone bed shear-stress $\tau_c \equiv \tau_0$, for rough or smooth beds as appropriate.
2. Use the *Procedure* in Section 3.3 to calculate the wave-alone bed shear-stress $\tau_w$, for rough or smooth beds as appropriate.
3. If the bed is rough (e.g. has worm-casts or fine drainage channels), use Equations (3.25), (3.26) and (3.27) to calculate $\tau_m$, $\tau_{max}$ and $\tau_{rms}$ respectively. This method gives the *total* bed shear-stresses experienced by the flow, including form drag on the roughness elements, which will control the current speed and determine the turbulence intensity.
4. If the bed is smooth (e.g. freshly deposited mud), use Equations (3.28), (3.26) and (3.27) to calculate $\tau_m$, $\tau_{max}$ and $\tau_{rms}$ respectively. This method gives the skin-friction shear-stresses acting on the mud, which are the relevant quantities for comparing with thresholds of erosion and deposition and for calculating erosion and deposition rates.
5. Alternatively, use the theoretical model and computational procedure of Myrhaug and Slaattelid (1990).

*Example 3.3. Bed shear-stress for combined waves and currents*
Calculate the mean, maximum and RMS bed shear-stress if the current in Example 3.1 is combined at 90° with the waves in Example 3.2. The current was smooth turbulent flow and the waves were laminar flow, so use the smooth-bed combined expression given by Equation 3.28 to calculate $\tau_m$.

| | | |
|---|---|---|
| From Example 3.1, current-alone bed shear-stress | $\tau_c$ | $0{\cdot}276\,\mathrm{N\,m^{-2}}$ |
| From Example 3.2, wave-alone bed shear-stress | $\tau_w$ | $0{\cdot}465\,\mathrm{N\,m^{-2}}$ |
| Calculate $\tau_m$ from Equation (3.28) mean bed shear-stress | $\tau_m$ | $0{\cdot}313\,\mathrm{N\,m^{-2}}$ |
| Calculate $\tau_{max}$ from Equation (3.26), maximum bed shear-stress | $\tau_{max}$ | $0{\cdot}561\,\mathrm{N\,m^{-2}}$ |
| Calculate $\tau_{rms}$ from Equation (3.27), RMS bed shear-stress | $\tau_{rms}$ | $0{\cdot}455\,\mathrm{N\,m^{-2}}$ |

# Erosion

# 4. Erosion

## 4.1. GENERAL

The threshold bed shear-stress for erosion $\tau_e$ is an important factor in many engineering studies. The threshold for erosion determines whether sediment deposited on the bed of an estuary or in the sea can be eroded by the prevailing current and wave conditions. Once the threshold has been exceeded then the rate at which sediment is eroded from the bed becomes the important factor determining how much erosion of the bed takes place in a particular time or by how much the sediment concentration will increase in the water column.

## 4.2. EROSION BY CURRENTS

### *Knowledge*

The flocs on the surface of a cohesive sediment bed are bound together by interparticle attractive forces. To remove a floc by flowing water requires a bed shear-stress $\tau_0$ sufficient to overcome the attractive forces. The critical erosion shear-stress $\tau_e$ of a cohesive sediment surface is defined as the shear stress required to be exerted by the flowing water to cause erosion of flocs.

The erosion shear-stress $\tau_e$ ($\mathrm{N\,m^{-2}}$) at any depth in a recently formed, partially consolidated cohesive sediment-bed was shown by laboratory studies (Thorn and Parsons, 1980) to be related to the dry density $C_M$ by the following form of empirical equation

$$\tau_e = E1\, C_M{}^{E2} \qquad \text{SC}\,(4.1)$$

where *E1* and *E2* are site-specific dimensional coefficients (SI units). For tests on mud from Brisbane, Grangemouth and

Belawan (see Table 4 for mud characteristics) Thorn and Parsons determined $E1 = 5{\cdot}42 \times 10^{-6}$ and $E2 = 2{\cdot}28$. These coefficients relate to freshly deposited beds in saline water (26–30 ppt) that have settled out and consolidated to a thickness of about 50 mm after 2–3 days. The relationship of Thorn and Parsons is valid in the range of dry density between 30–200 kg m$^{-3}$. Alternative values derived from a larger dataset, including those used by Thorn and Parsons, are $E1 = 0{\cdot}0012$, $E2 = 1{\cdot}2$ (Delo and Ockenden, 1992) valid in the range 30–400 kg m$^{-3}$ (Figure 13).

Recent research by Mitchener *et al.* (1996b) has extended the relationship to cover a wider range of cohesive bed types: mixed beds, consolidated beds and blended high-density beds up to bulk densities of 1800 kg m$^{-3}$ (Figure 14).

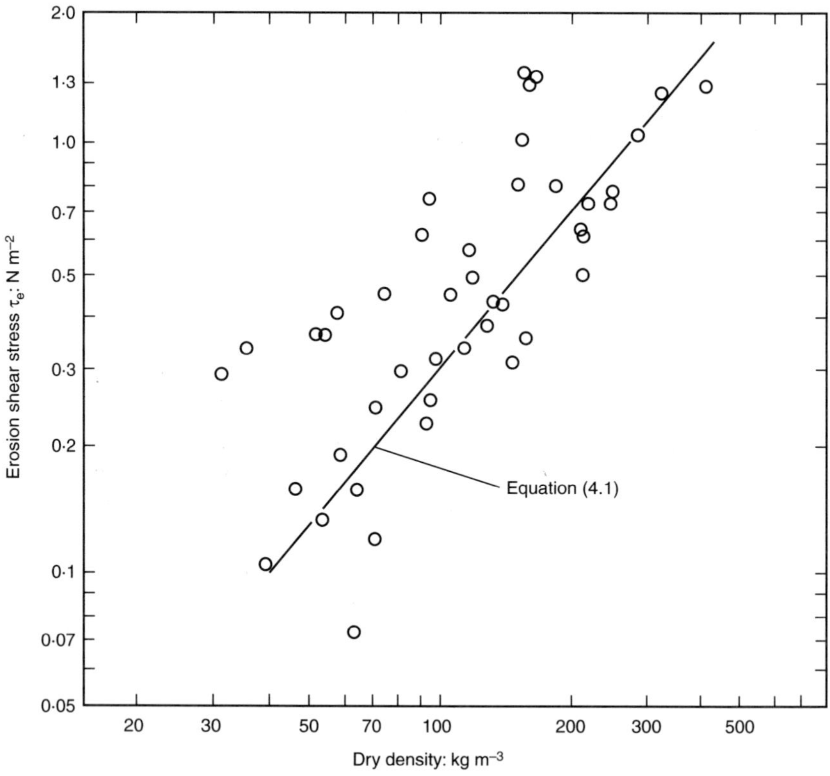


*Figure 13. Erosion threshold $\tau_e$ as a function of bed dry density $C_M$ obtained in laboratory tests (from Delo and Ockenden, 1992)*

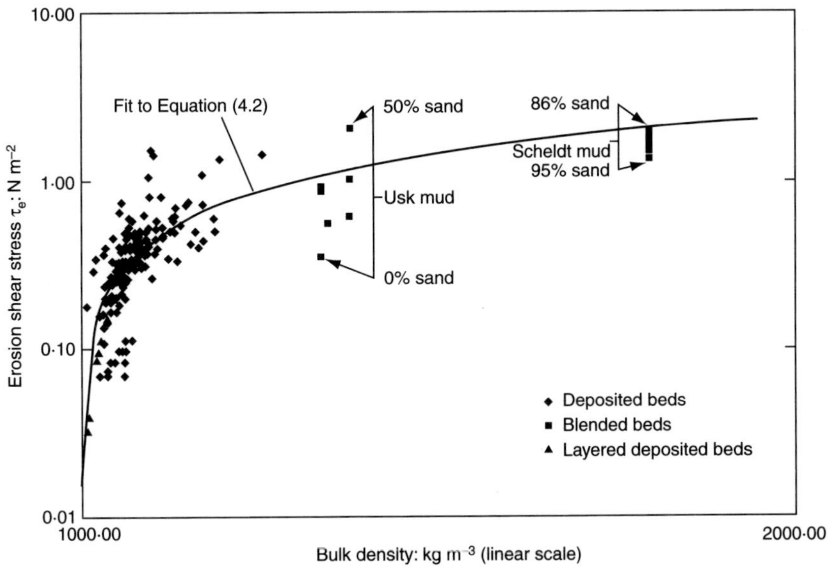


*Figure 14. Erosion threshold $\tau_e$ for cohesive sediments as a function of bed bulk density $\rho_B$ from data for partially consolidated, redeposited beds obtained in laboratory tests (after Mitchener* et al., *1996a)*

The relationship is of the form:

$$\tau_e = E3(\rho_B - 1000)^{E4} \qquad \text{SC}(4.2)$$

where $\rho_B$ is the bulk density of the bed, and *E3* and *E4* are dimensional coefficients (SI units) fitted to the data with $E3 = 0{\cdot}015$ and $E4 = 0{\cdot}73$ (Mitchener *et al.*, 1996b).

The above relationships are derived from laboratory tests with artificially deposited beds of natural muds. The majority of these tests were carried out in the large annular flume (Carousel) at HR Wallingford (Ockenden and Delo, 1991). The value of laboratory experiments is that they are carried out under controlled conditions, although these may not reflect the full environmental conditions found in natural estuarine muds. Amos *et al.* (1998) compared equations (4.1) and (4.2) against in-situ sea bed measurements taken with their Sea Carousel (Amos *et al.*, 1992). Although there was considerable scatter in their data, a consistent correlation between $\rho_B$ and $\tau_e$ was obtained and the measured values of $\tau_e$ were considerably smaller than are given

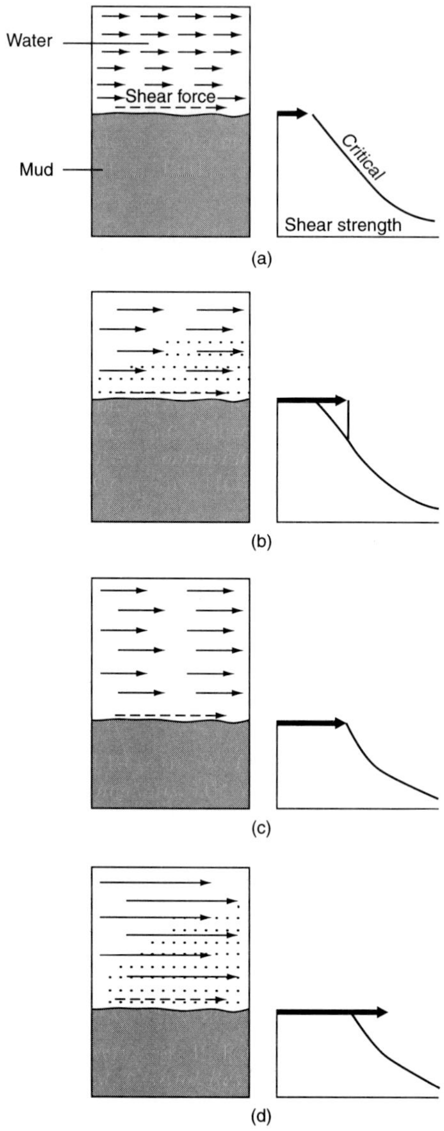


*Figure 15. Schematic diagram illustrating the cohesive sediment erosion process: (a) no erosion; (b) erosion starts when shear-stress exceeds shear-strength at surface; (c) erosion stops when exposed mud has a shear strength equal to or greater than the applied stress; (d) erosion restarts with increased velocity giving higher shear-stresses than the strength of the exposed layer (from Delo and Ockenden, 1992)*

by Equations (4.1) and (4.2). Typically $\tau_e$ did not exceed $1\,\mathrm{N\,m^{-2}}$ over the density range $1000 < \rho_B(\mathrm{kg\,m^{-3}}) < 2000$. A similar result was found by Mitchener *et al.* (1996b) from their in-situ data. Values of the erosion shear-stress $\tau_e$ measured in-situ on intertidal sediments are in the range $0{\cdot}02$–$5{\cdot}0\,\mathrm{N\,m^{-2}}$, typically around $0{\cdot}1$–$0{\cdot}2\,\mathrm{N\,m^{-2}}$ (Mitchener, 1996).

The erosion shear-stress of a cohesive sediment bed usually increases with depth together with the dry density. Therefore, under a constant fluid bed shear-stress $\tau_0$ an eroding cohesive sediment bed will eventually cease to erode when the erosion strength of the exposed cohesive sediment $\tau_e$ is equivalent to $\tau_0$ (Figure 15).

At shear stresses well in excess of the critical floc erosion shear-stress, a cohesive sediment bed may experience mass erosion. The process of mass erosion comprises the detachment of lumps of cohesive sediment from the bed. There is little quantitative field data or experimental data on the mass erosion of cohesive sediment.

The rate of erosion of cohesive sediment has been studied in laboratory flume experiments (HR Wallingford, 1989; Mitchener *et al.*, 1996a) and to a lesser extent in the field (Diserens *et al.*, 1991; Amos *et al.*, 1992; Amos *et al.*, 1998). It has been found that the rate of erosion $dm/dt$, expressed as dry mass of mud eroded per unit area per unit time ($\mathrm{kg\,m^2\,s^{-1}}$), is related to the magnitude of the excess shear-stress $(\tau_0 - \tau_e)$ by the dimensional coefficient $m_e$, known as the erosion constant. This can be expressed using the empirical equation based on the work of Partheniades (1965), see Owen (1977):

$$dm/dt \begin{cases} = m_e(\tau_0 - \tau_e) & \text{for } \tau_0 > \tau_e \\ = 0 & \text{for } \tau_0 \leqslant \tau_e \end{cases} \qquad \text{SC (4.3)}$$

Figure 16 shows Equation (4.3) – solid line – with $m_e = 0{\cdot}002\,\mathrm{kg\,N^{-1}\,s^{-1}}$ or $\mathrm{m\,s^{-1}}$. However, there is considerable variation in the erosion properties of cohesive sediments from different sites as indicated by the range of experimental scatter about the average relationship in Figure 16 for the erosion properties of cohesive sediments tested at HR Wallingford. Table 8 lists typical values and ranges of the erosion parameters for a large number of different muds (Table 9) tested at HR Wallingford.

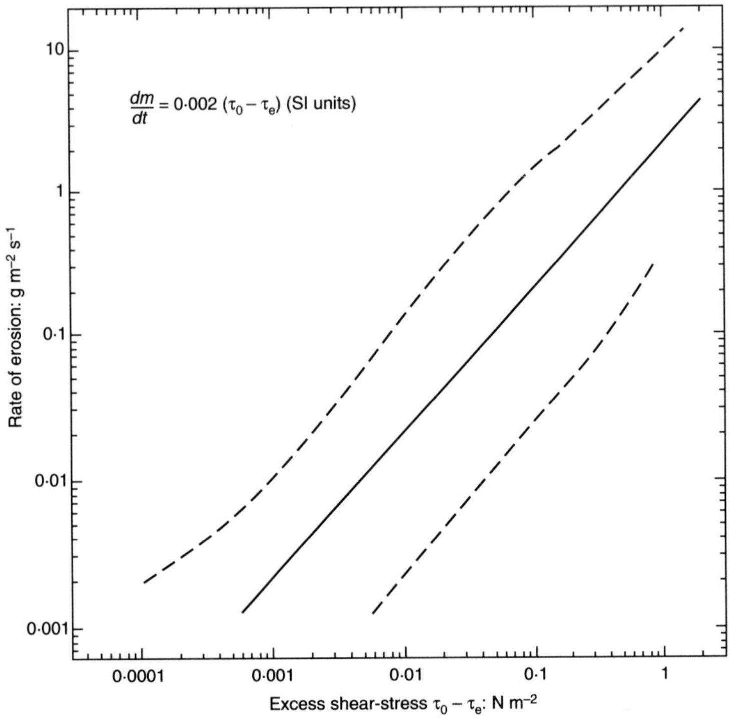


*Figure 16. Rate of erosion in currents as a function of excess bed shear-stress (from Delo, 1988a). The dashed lines indicate the limits of scatter in the experimental data*

## Procedure

If at all possible it is recommended to measure the erodibility of muddy sediments as they occur in situ, because the effects of mud collection and storage can alter significantly the sediment behavioural characteristics (Williamson and Ockenden 1993, Berlamont *et al.*, 1993). For some natural muddy sediments the biological, chemical and environmental conditions may change the erodibility parameters by more than an order of magnitude. If no new data are available then use representative results from previous studies.

### *Field*

1. Use an in-situ erodibility instrument to measure the erosion parameters directly under field conditions. HR Wallingford

*Table 8. Comparison of muds tested at HR Wallingford – erosional/depositional properties determined in laboratory tests*

| Mud type | Erosion constant $m_e$: $kg\,N^{-1}\,s^{-1}$ | Erosion shear stress $\tau_e$: $N\,m^{-2}$ | Permeability $K_I$: $m\,s^{-1}$ | Effective stress $\sigma'$: $N\,m^{-2}$ | Settling velocity $w_{50}$: $m\,s^{-1}$ | Erosion shear-stress at $C_M = 50\,kg\,m^{-3}$: $N\,m^{-2}$ | Shear-stress at field $C_M$: $N\,m^{-2}$ | Permeability $C_M = 100\,kg\,m^{-3}$: $m\,s^{-1}$ |
|---|---|---|---|---|---|---|---|---|
| Cardiff Taff/Ely | – | $\tau_e = 0{\cdot}00022\,C_M^{1{\cdot}5}$ | $K_I = 10^{-9}\,V_r^2$ | $\sigma' = 3000\,V_r^{-2} - 4{\cdot}5$ | $w_{50} = 0{\cdot}0003\,C_M^{1{\cdot}0}$ | 0·08 | 2·5 | $6 \times 10^{-6}$ |
| Cardiff Rhymney | – | $\tau_e = 0{\cdot}0042\,C_M^{0{\cdot}9}$ | $K_I = 10^{-9}\,V_r^2$ | $\sigma' = 4000\,V_r^{-2} - 4{\cdot}0$ | $w_{50} = 0{\cdot}0003\,C_M^{1{\cdot}0}$ | 0·14 | 1·1 | $6 \times 10^{-6}$ |
| Fawley | – | $\tau_e = 0{\cdot}0005\,C_M^{1{\cdot}43}$ | $\log(K_I) = -0{\cdot}014\,C_M - 4{\cdot}9$ | $\sigma' = 0{\cdot}004\,C_M^2 +0{\cdot}058\,C_M + 7{\cdot}6$ | $w_{50} = 0{\cdot}0075\,C_M^{1{\cdot}9}$ | 0·13 | 2·2 | $5 \times 10^{-7}$ |
| Grangemouth | 0·0005–0·0014 | $\tau_e = 0{\cdot}0045\,C_M^{0{\cdot}9}$ | $\log(K_I) = -0{\cdot}01\,C_M - 5{\cdot}5$ | $\sigma' = 0{\cdot}0005\,C_M^2 -0{\cdot}01\,C_M + 0{\cdot}05$ | – | 0·15 | 1·4 | $3 \times 10^{-7}$ |
| Harwich | 0·0007 | $\tau_e = 0{\cdot}00035\,C_M^{1{\cdot}43}$ | $\log(K_I) = -0{\cdot}0115\,C_M - 5{\cdot}3$ | $\sigma' = 0{\cdot}0011\,C_M^2 -0{\cdot}15\,C_M + 15{\cdot}0$ | $w_{50} = 0{\cdot}002\,C_m^{2{\cdot}0}$ | 0·09 | 1·8 | $4 \times 10^{-7}$ |
| Hong Kong | 0·0006–0·0015 | $\tau_e = 0{\cdot}0013\,C_M^{1{\cdot}2}$ | – | – | – | 0·14 | – | – |
| Ipswich | 0·0009–0·0030 | $\tau_e = 0{\cdot}00028\,C_M^{1{\cdot}55}$ | – | – | $w_{50} = 0{\cdot}0012\,C_M^{1{\cdot}4}$ | 0·12 | 4·4 | – |
| Kelang | 0·0002–0·0009 | $\tau_e = 0{\cdot}0005\,C_M^{1{\cdot}4}$ | – | – | – | 0·12 | – | – |
| Kingsnorth | 0·0007 | $\tau_e = 0{\cdot}005\,C_M^{0{\cdot}9}$ | $\log(K_I) = -0{\cdot}017\,C_M - 4{\cdot}75$ | $\sigma' = 0{\cdot}0009\,C_M^2 -0{\cdot}09\,C_M + 2{\cdot}25$ | $w_{50} = 0{\cdot}0014\,C_M^{1{\cdot}0}$ | 0·17 | 1·6 | $4 \times 10^{-7}$ |
| Medway | 0·0007 | $\tau_e = 0{\cdot}0007\,C_M^{1{\cdot}3}$ | $K_I = 10^{-10}\,V_r^3$ | $\sigma' = 1600\,V_r^{-1{\cdot}8} - 2{\cdot}3$ | – | 0·11 | 1·4 | $1{\cdot}6 \times 10^{-5}$ |
| Mersey Eastham | 0·0005 | $\tau_e = 0{\cdot}013\,C_M^{0{\cdot}7}$ | $\log(K_I) = -0{\cdot}015\,C_M - 4{\cdot}25$ | $\sigma' = 0{\cdot}0007\,C_M^2 -0{\cdot}07\,C_M + 1{\cdot}75$ | – | 0·20 | 0·8 | $2 \times 10^{-6}$ |
| Mersey Runcorn | – | – | $\log(K_I) = -0{\cdot}01\,C_M - 4{\cdot}25$ | $\sigma' = 0{\cdot}0004\,C_M^2 -0{\cdot}04\,C_M + 1{\cdot}0$ | – | – | – | $6 \times 10^{-6}$ |
| Poole | 0·0007–0·0014 | $\tau_e = 0{\cdot}0003\,C_M^{1{\cdot}5}$ | – | $\sigma' = 9000\,V_r^{-3} - 0{\cdot}204$ | – | 0·11 | 6·8 | – |
| Tees Seal Sands | 0·0002–0·0014 | $\tau_e = 0{\cdot}0025\,C_M^{1{\cdot}0}$ | – | – | – | 0·13 | 2·2 | – |
| Tees dredged | 0·0005–0·0018 | $\tau_e = 0{\cdot}00014\,C_M^{1{\cdot}7}$ | * $\log K_I = -0{\cdot}011\,C_M - 5{\cdot}0$ | * $\sigma' = 0{\cdot}0008\,C_M^2 -0{\cdot}016\,C_M + 0{\cdot}08$ | – | 0·11 | 9·6 | $8 \times 10^{-7}$ |

Note. $C_M$ = dry density in $kg\,m^{-3}$, $V_r$ = voids ratio, ($e = [2650/C_M] - 1$)
* relationship is for beds 70–80 mm final thickness

*Table 9. Characteristics of the muds in Table 8*

| Mud type | % Silt: ⩽63 microns | Organics: % | Cation Exchange Capacity: meq/100 g | Mineralogy | Bulk density: $kg\,m^{-3}$ |
|---|---|---|---|---|---|
| Cardiff Taff/Ely | 80–100 | 12·6–14·9 | – | – | 1300–1850 |
| Cardiff Rhymney | >90 | 10·6–12·8 | – | – | 1300–1850 |
| Fawley | 95 | 2·2–2·4 | 23·6–24·2 | Quartz 26% Clays 74% | 1220–1650 |
| Grangemouth | 80–90 | 4·7–5·5 | 18·8–20·3 | Quartz 18% Clays 82% | 1370 |
| Harwich | 88–95 | 2·2–2·7 | 21·5–23·6 | Quartz 24% Clays 76% | 1250 |
| Hong Kong | 65–80 | 6·9–9·6 | 14·0 | Quartz 20% Clays 80% | – |
| Ipswich | – | – | – | Quartz 17% Clays 83% | 1320 |
| Kelang | 65–80 | – | – | Quartz 17% Clays 83% | – |
| Kingsnorth | 64 | 4·8 | 26·0 | Quartz 17% Clays 83% | 1375 |
| Medway | 80 | 2·7–2·9 | 20·9–22·9 | Quartz 21% Clays 79% | 1220 |
| Mersey Eastham | 80 | 8·6 | 14·1 | – | 1140 |
| Mersey Runcorn | 45 | 6·3 | 17·2 | – | 1250 |
| Poole | 80–85 | 2·2–2·3 | 20·6–20·7 | Quartz 30% Clays 70% | 1500 |
| Tees Seal Sands | 75 | 4·0–4·5 | 15·1–15·7 | Quartz 19% Clays 81% | 1550 |
| Tees Dredged | 75 | 10·1–11·1 | 17·6–18·1 | Quartz 22% Clays 78% | 1430 |

developed SedErode – Sediment Erosion Device (Mitchener *et al.*, 1996b), which is a field instrument that allows easy measurement of the critical erosion shear-stress in situ on natural muddy sediments. There are also other types of erodibility instruments for different laboratory and field applications available, as reviewed by Mitchener (1996) and Black and Paterson (1997). Depending on the scope of the study either correlate the erosion parameters with relevant sediment properties or calculate the average value and its variability.

SandCalc uses a default value for $m_e$ of $0{\cdot}001\,\mathrm{kg\,N^{-1}\,s^{-1}}$ in Equation (4.3). Where near-bed suspended sediment data are available check the erosion predictions against the observed time series of concentration.

*Laboratory*

2. Obtain a sample of cohesive sediment from the site and investigate its erosion properties in a laboratory flume (e.g. Ockenden and Delo, 1991). Wherever possible, maintain the natural bed surface and sediment structure. Derive the values of the constants in Equations (4.1) or (4.2) and (4.3) from the experimental results.

*Existing field or laboratory measurements*

3. If the site of interest has been investigated in the past by field or laboratory experiments and the results are still appropriate use those values (as step 1).

*No field or laboratory measurements*

4. Use with caution the average values of the erosion parameters for Equations (4.1) or (4.2) and (4.3) shown in Figures 13, 14 and 16. Table 8 also lists typical values for $\tau_e$ and $m_e$.

*Example 4.1. Erosion by currents*

To calculate the depth of erosion of a mud bed deposited at the bed of an estuary, if the current and other prevailing conditions given in Example 3.1 act on the bed for one hour, obtain, additionally, an estimate of:

| | | |
|---|---|---|
| dry density of mud | $C_M$ | $70\,\mathrm{kg\,m^{-3}}$ |

Calculate the threshold bed shear-stress for erosion from Equation (4.1). No site-specific values of coefficients *E1* and *E2* are available in this example so use default values $E1 = 0{\cdot}0012$ and $E2 = 1{\cdot}2$ in SI units (Delo and Ockenden, 1992), or use SandCalc–Mud–Erosion–Threshold–Thorn and Parsons

| | | |
|---|---|---|
| erosion shear-stress | $\tau_e$ | $0{\cdot}197\,\mathrm{N\,m^{-2}}$ |

Calculate the rate of erosion from Equation (4.3), using the value of $\tau_0 = 0{\cdot}276\,\mathrm{N\,m^{-2}}$ calculated in Example (3.1) and the default value $m_e = 0{\cdot}001\,\mathrm{kg/Ns}$, or use SandCalc–Mud–Erosion–Currents–Partheniades

| | | |
|---|---|---|
| erosion rate | $dm/dt$ | $7{\cdot}94 \times 10^{-5}\,\mathrm{kg\,m^2\,s^{-1}}$ |

Over one hour the mass eroded per square metre is

| | | |
|---|---|---|
| $3600 \times 7{\cdot}94 \times 10^{-5}$ | $m$ | $0{\cdot}286\,\mathrm{kg\,m^{-2}}$ |
| The depth of bed eroded over one hour is thus | $m/C_M$ | $0{\cdot}00408\,\mathrm{m}$ |

Thus the deposited bed is eroded by the given current at 4·1 mm per hour.

## 4.3. EROSION BY WAVES

### *Knowledge*

Consideration of the physical processes of waves as an eroding mechanism of mud is complicated by the complexity of the mud bed response. Changes in the characteristics of surface waves due to mud motion, wave energy dissipated in the mud and the erosion of the mud are all interlinked. Hence, the estimation of shear stress at the bed surface for the purpose of correlation with the rate of erosion should not necessarily be based on the assumption of a rigid bed (Maa and Mehta, 1985).

The action of the waves may be sufficient to cause surface erosion, with material passing directly into suspension. Experiments with a natural mud bed have shown that wave action erodes mud of a given dry density at about the same peak shear-stress as that required with unidirectional flow. The erosion rate of a mud bed at $\rho_B = 1280\,\mathrm{kg\,m^{-3}}$ was found to be similar to the proportional excess shear-stress relationship for current induced erosion given in Equation (4.3) (Diserens and Delo, 1988).

To calculate the mean erosion rate, $dm/dt$, of a mud bed by waves, the steady-current bed shear-stress $\tau_0$ in Equation (4.3) is replaced by the maximum bed shear-stress $\tau_w$ during a wave-cycle:

$$dm/dt \begin{cases} = m_e(\tau_w - \tau_e) & \text{for } \tau_w > \tau_e \\ = 0 & \text{for } \tau_w \leqslant \tau_e \end{cases} \qquad \text{SC (4.4)}$$

A very high proportion of the eroded mud will be contained within the relatively thin wave boundary layer because the sharp-density gradient damps the vertical turbulent exchange with the water column above. The vertical turbulent exchange and transport rate increases as the flow velocity of the overlying water increases.

## *Procedure*

The following approach can be taken when predicting erosion by waves.

1. Conduct laboratory tests using site specific mud samples to determine the critical wave shear-stress and the erosional characteristics of higher density beds.

   Or:

2. For an order of magnitude, estimate of erosion rate use Equation (4.4). Methods of calculating $\tau_w$ for smooth or rough bed conditions are described in Section 3.3. Values for $\tau_e$ can be calculated from Equation (4.1) or (4.2), or Table 8 can be used. SandCalc uses a default value for $m_e$ of $0{\cdot}001\,\mathrm{kg\,N^{-1}\,s^{-1}}$. Where near-bed suspended sediment data is available check the erosion predictions against the observed time series of concentration.

Following the same reasoning that the form of the erosion rate in waves is similar to that found under currents-alone it may be possible to use Equation (4.4) to predict the erosion in combined wave and current flow. It is most appropriate to replace the value of $\tau_w$ with the value of $\tau_{max}$ calculated in Section 3.4. However, in the Authors' experience, there are no data available to support the form of such an expression. In combined wave and current flow the vertical turbulent exchange in the current boundary layer will facilitate the mixing of eroded sediment into the water column.

*Example 4.2. Erosion by waves*

Calculate the depth of erosion of the mud bed specified in Example 4.1 if it is acted on by the waves specified in Example 3.2 for one hour, instead of the current.

Use Equation (4.4) to calculate the rate of erosion, using the value of $\tau_w = 0{\cdot}465\,\mathrm{N\,m^{-2}}$ calculated in Example 3.2, the value of $\tau_e = 0{\cdot}197\,\mathrm{N\,m^2}$ calculated in Example 4.1, and the default value of $m_e = 0{\cdot}001\,\mathrm{kg\,N^{-1}\,s^{-1}}$, or use SandCalc–Mud–Erosion–Waves.

| | | |
|---|---|---|
| erosion rate | $dm/dt$ | $2{\cdot}68 \times 10^{-4}\,\mathrm{kg\,m^2\,s^{-1}}$ |

Over one hour the mass eroded per square metre is

| | |
|---|---|
| $3600 \times 2{\cdot}68 \times 10^{-4} =$ | $0{\cdot}965\,\mathrm{kg\,m^{-2}}$ |

The depth of bed eroded over one hour is thus

| | |
|---|---|
| $m/C_M$ | $0{\cdot}0138\,\mathrm{m}$ |

Thus the wave action erodes the bed at 13·8 mm per hour.

# Suspension of mud in the water column

# 5. Suspension of mud in the water column

## 5.1. GENERAL

Once the mud has been eroded the flocs and constituent particles are transferred away from the bed into the water column by turbulent mixing (Figure 3). The concentration of material in the water column will determine how the suspension behaves and how it is transported.

## 5.2. PROPERTIES OF MUD SUSPENSIONS

### *Knowledge*

The main mode of transport of mud is as a suspension; it remains in suspension for long periods and travels long distances before depositing. However, the behaviour of suspended mud is complex and varied. Hence the study of suspended mud has been a dominant theme of cohesive sediments research.

The behaviour of a mud suspension depends primarily on its local concentration, and this has been used by various researchers to categorise suspended mud into a number of classes. At any particular location several of these categories may exist overlying one another in layers, with the lowest concentrations in the upper part of the water column, and the highest concentrations in the bed. Kirby and Parker (1973, 1983) identified three states of mud:

- mobile suspensions, with a mass concentration between 1 and $150\,\mathrm{kg\,m^{-3}}$
- stationary suspensions – settling but not moving, with a mass concentration between 0·1 and $200\,\mathrm{kg\,m^{-3}}$

- settled mud – consolidating and not moving, with a bulk density between 1300 and 1700 kg $m^{-3}$.

The EU research project COSINUS has proposed the following categories, on the basis of the local concentration, its vertical gradient, and the flow speed. Taken together these form the gradient Richardson number, $Ri_g$, which characterises the turbulence structure of the flow and is defined by:

$$Ri_g = \frac{-g\,\partial\rho/\partial z}{\rho(\partial U/\partial z)^2} \tag{5.1}$$

where g = acceleration due to gravity
$z$ = height above bed (positive upwards)
$\rho$ = water density including sediment at height $z$
$U$ = current speed at height $z$

Small values of $Ri_g$ indicate that the production of turbulence by shear greatly exceeds the stabilising effect of the density gradient, and hence the flow is fully turbulent. Flows with $Ri_g < 0{\cdot}03$ have a negligible stratification. The effect of turbulence damping due to density stratification increases with $Ri_g$ until, at a value somewhere between 0·1 and 0·3, the density gradient is so strong that the turbulence is damped out and the flow becomes effectively laminar.

The COSINUS categories are as follows.

### *Dilute suspension*

Characterised by low values of $Ri_g$, so usually fully turbulent. Typical dry mass concentrations are less than 1 kg $m^{-3}$.

### *Concentrated benthic suspension (CBS)*

Significant values of $Ri_g$, so a significant amount of stratification. Concentration low enough that behaviour is still Newtonian with a viscosity similar to that of water. Sediment still kept in suspension by turbulence: CBS is turbulent although water column above it can be laminar in principle, if strongly stratified with turbulence production at the bed only. Typical dry mass concentrations are less than 10 kg $m^{-3}$.

*Fluid mud*
This occurs when the mud flocs become space filling. It is typically non-Newtonian often with a high viscosity. The flow is usually laminar. A typical mode of formation is the collapse of a CBS, when the energy input is no longer sufficient to maintain turbulence. The rate of dewatering of fluid mud is limited by the high volume concentration.

*Consolidating bed*
Formed either from settling from suspension or from the formation of a structure and hence a degree of strength in a fluid mud layer. If a fluid mud layer remains undisturbed, bonds will form between the flocs. The self-weight of the structure will slowly expel the pore water.

Winterwerp (1999) elaborated on the COSINUS categories as follows.

*Low-concentrated mud suspension*
This is a suspension of cohesive sediment at a concentration of the order of several 10 mg/l to maybe a few 100 mg/l. This suspension behaves Newtonian and does not significantly affect the turbulent flow field.

*High-concentrated mud suspension (HCMS)*
HCMS is a suspension of cohesive sediment at a concentration of a few 100 mg/l to a few 1000 mg/l. This suspension behaves Newtonian, though possibly with a somewhat increased viscosity, and is transported with the main flow. An important feature of HCMS is its interaction with the turbulent flow field.

*Fluid mud*
This is a suspension of cohesive sediment at a concentration beyond the gelling point, i.e. of the order of several 10 g/l to 100 g/l. This suspension exhibits profound non-Newtonian behaviour, and it is either stationary or moving. In the latter case the fluid mud flow may be laminar or turbulent and its dynamics will be fairly independent of the flow in the water column above.

Concentrations between those typical for fluid mud and HCMS do occur in practice, but they are transient and the suspension cannot be stable and no intermediate category is required.

Winterwerp's three categories convert to ranges of (mass) concentration measured in kg m$^{-3}$ as: 0·05 to 0·3 kg m$^{-3}$, 0·3 to 3 kg m$^{-3}$, and 50 to 100 kg m$^{-3}$, respectively, if we interpret 'several' as 5 and 'a few' as 3.

Soulsby (2000) categorised the behaviour of suspended mud in terms of its settling behaviour as follows.

### *Unflocculated suspension*

Known as *washload* in rivers, where the fresh water ensures that electrostatic repulsion between individual grains is stronger than the molecular attractive force, so the grains stay separate (see Section 2.4.2). Also found in the sea where concentrations are sufficiently low (<0·01 kg m$^{-3}$) that grains do not collide to form flocs. The settling velocity of the grains is of order 0·01 mm s$^{-1}$. The time taken for a grain to settle through 10 m is about 12 days.

### *Flocculated suspension*

Found in estuaries at salinities greater than about 3 ppt and and in the sea at mass (dry) concentrations greater than about 0·01 kg m$^{-3}$, up to a dry mass concentration of about 3 kg m$^{-3}$. Grains collide through differential settling and turbulent shear. Floc size increases with concentration, as there is a greater probability of collisions. The flocs behave like low-density particles whose size can vary with concentration, but which do not interact with each other very strongly. The volume-concentration of grains within the flocs is typically about 0·03, corresponding to an internal floc density of about 1080 kg m$^{-3}$ in water of density 1027 kg m$^{-3}$. The flow velocity of water between the grains in the floc is much smaller than the flow velocity of the floc through the surrounding water. Settling velocities of the flocs are in the range 0·01 to 5 mm s$^{-1}$, and the time taken to settle through 10 m is in the range 30 minutes to 12 days.

### *Fluid mud*

Found in patchy pools at the bottom of estuaries (and possibly the sea), with mass (dry) concentrations between about 3 kg m$^{-3}$ and 100 kg m$^{-3}$. The concentration is large enough that the gap between flocs is similar to the size of the flocs, and hence the flocs interact with each other so that hindered settling occurs. The variation of settling velocity with concentration reaches a maximum value in the range 1 to $10\,\mathrm{mm\,s^{-1}}$ at a concentration somewhere between 3 and $30\,\mathrm{kg\,m^{-3}}$. At higher concentrations than this, the settling velocity *decreases* strongly with concentration to become effectively zero at about 50 to $100\,\mathrm{kg\,m^{-3}}$. The viscosity of the mud is much higher than that of pure water, but sufficiently low that it is easily mobilised. See Sections 6.3, 6.4 and 6.5.

### *Consolidating settled bed*

At mass (dry) concentrations in excess of about 50 to $100\,\mathrm{kg\,m^{-3}}$ the viscosity is sufficiently high that the mud does not flow, although it could move as a creeping, viscous flow down a sufficiently steep slope (Section 7.3). The flocs are touching each other and progressively collapsing under the overlying weight, until the grains themselves are touching. The flow velocity of water through the flocs exceeds the flow past the flocs (see Section 5.3).

The various categorisation schemes all provide insight into the behaviour of mud suspensions (including the consolidating bed) in complementary ways. The border-lines between the categories differ by up to a factor of 3, but in practice these border-lines will not be distinct, and will also depend on the mineralogy of the grains and the salinity of the water. A range of definitions encountered in the literature are depicted in Figure 17 with their quoted range in concentration.

The concentration of mud in suspension can be controlled by the density stratification it produces, giving a maximum or *saturated* concentration for a particular flow and sediment condition. Winterwerp *et al.* (2000) investigated this phenomenon. Experiments have shown that turbulence in a flow collapses when the flux Richardson number exceeds a critical value. Experiments involving fresh and saline water have established a critical value of approximately 0·15. The flux Richardson number, $Ri_f$, is another stability parameter, like the gradient and bulk Richardson numbers (Equations (5.1) and (6.5)) to which it is related, that is used to determine the effect of density stratification on a flow. It is defined by:

$$Ri_f = \frac{gF_D}{\tau(\partial U/\partial z)} \tag{5.2}$$

where $F_D$ is the upward flux of density (sediment), and $\tau$ is the shear-stress at height $z$. The choice of stability parameter used

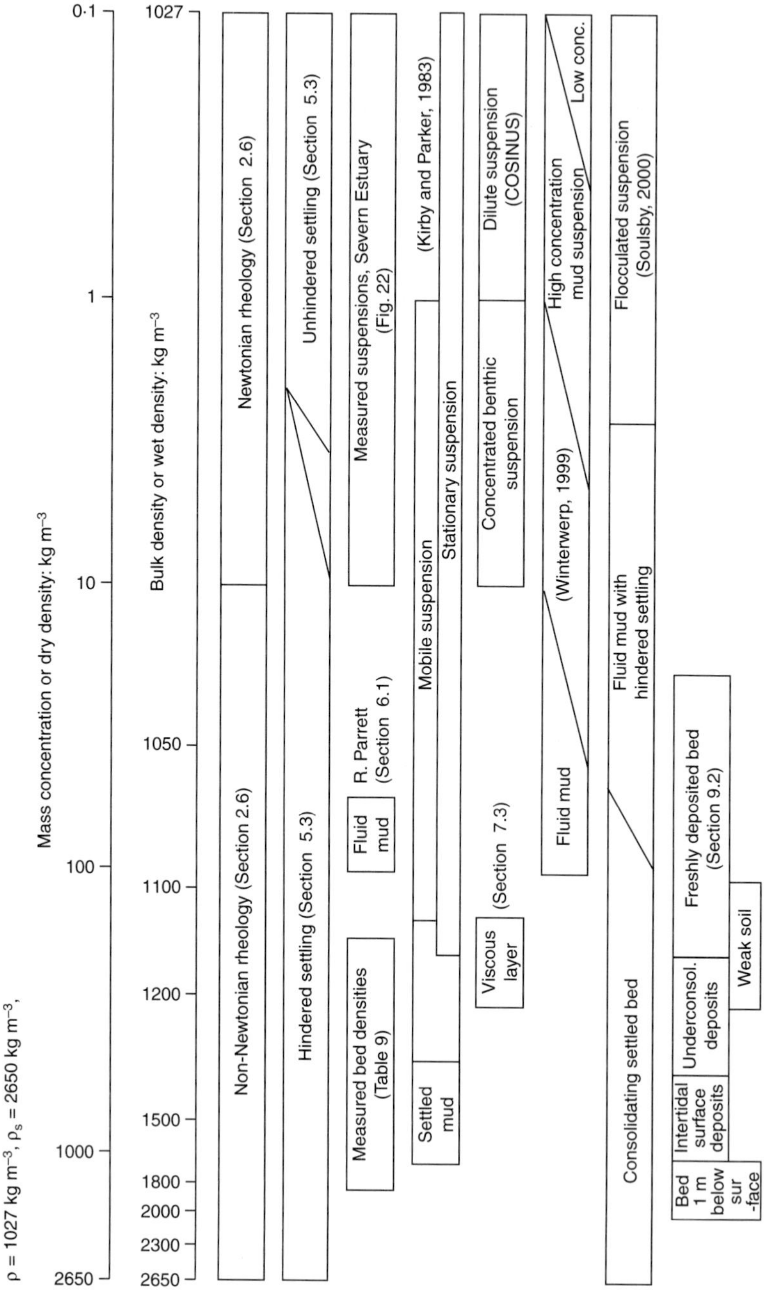


*Figure 17. Definitions for mud suspensions found in the literature*

depends partly on the physical situation, but is also partly a matter of preference.

By calculating the vertical profile of suspended sediment, based on the usual balance of settling and mixing fluxes, the vertical density gradient can be calculated. This gives an indication of how much concentration in suspension leads to the critical value of Richardson number. Depth-averaging of this concentration profile leads to the following equation of Winterwerp *et al.* (2000), applicable to steady flow conditions:

$$C_s = A_s \frac{\rho}{g} \frac{u_*^3}{h w_s} \tag{5.3}$$

where $C_s$ is the saturated dry mass depth-averaged concentration, $A_s$ is a dimensionless constant of proportionality and the other symbols have their usual meaning. Thus, if the flow is deep and slow, the saturated concentration can be quite low. The example presented in the Winterwerp paper involves a depth average velocity of 0·2 m s$^{-1}$, water depth of 20 m, settling velocity of 0·5 mm s$^{-1}$, giving a saturated concentration of about 0·024 kg m$^{-3}$. Winterwerp *et al.* give the value of $A_s$ as approximately 2·3. Saturated concentrations increase with the cube of the velocity, so a typical estuarine flow of 1 m s$^{-1}$ in 10 m of water, with a settling velocity of 1 mm s$^{-1}$ and a roughness value ($z_0$) of 0·003 m, gives a saturated concentration of approximately 4 kg m$^{-3}$.

In tidal environments, conditions tend to vary between well-mixed through the depth during the fastest flows, to the formation of lutoclines at times around slack water. The behaviour is complex and depends on the time scales of settling at slack water and re-mixing of fluid mud layer when the flow is fast, potentially leading to formation of multiple *lutoclines* – strong gradients in concentration – as observed by Kirby and Parker (1973, 1983) in the Severn Estuary (UK).

## 5.3. FLOCCULATION AND SETTLING VELOCITY

### *Knowledge*

Flocculation of sediment particles is the consequence of particles sticking together as they are brought into contact with each other. Collision and cohesion are the essential processes of flocculation (Krone, 1962).

The size and hence the settling velocity of a floc will be much larger than that of the constituent individual particles. The experimental curve in Figure 4 shows the ratio of the settling velocities of flocculated and chemically dispersed discrete particles (Migniot, 1968). For particles of diameter 0·1 micron the ratio is of the order $10^4$. For a diameter of 60 microns there is little flocculation and the settling velocity ratio is 1. However, in practice all estuarine sediments and suspended flocs have a range of constituent particle sizes and a simple settling velocity ratio does not exist.

The maximum floc size is governed by: the discrete constituent particle sizes, suspended sediment concentration, mineralogy, organic content, pH and ionic strength of the mud; the chemical composition of the pore water and suspending water; and the hydrodynamic parameters of the flow such as water temperature and salinity, the velocity and turbulence structure, internal shear and bed shear-stress.

Changes in salinity above 2–3 ppt have only a small affect on the settling velocity of cohesive sediment (Burt and Stevenson, 1983).

The size distribution of flocs (aggregates of discrete particles) in the field has been studied (Krone, 1972) and directly measured using a submerged laser particle analyser (Bale and Morris, 1987), submerged video recording of settling through an open-ended column (Van Leussen, 1988), and video image processing of settling in an Owen Tube (Dearnaley, 1991).

Measurement of the settling velocity of flocculated sediment is best done in the field as removal of a sample to the laboratory changes the size distribution of the flocs. The data shown in Figure 18 indicates that laboratory measurements of median settling velocity could be an order of magnitude lower than those measured in the field with an Owen Tube (Owen, 1971).

However, each Owen Tube measurement involves withdrawing samples from the tube over a period of one hour, and during this time the presence of the tube itself has an impact on the settling velocity. Video image analysis of settling in an Owen Tube has indicated that the median settling velocity determined by Owen Tube analysis may underestimate the actual settling velocity (Dearnaley, 1991). This underestimate happens

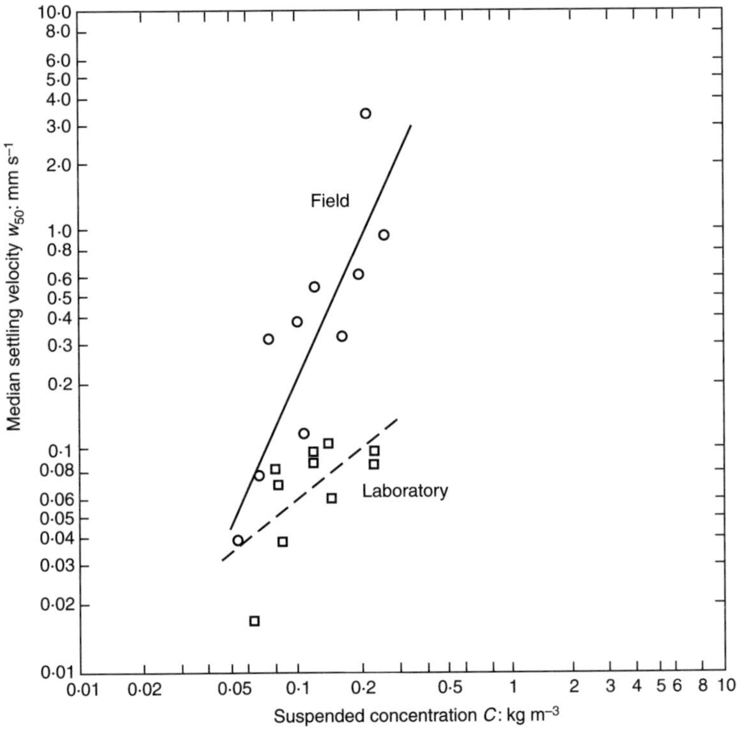


*Figure 18. Comparison of median settling velocities from the laboratory and the field (Owen Tube) plotted as a function of suspended sediment concentration (from Delo and Ockenden, 1992)*

because of the modifications that take place in the sampling chamber over the period of sampling. The Floc Camera and other in-situ devices provide a measure of the floc size and settling velocity soon after the sample has been taken.

Eight instruments for measuring floc settling characteristics in-situ have been compared in an experiment on the Elbe in 1993 (Dyer *et al.*, 1996; Eisma *et al.*, 1997). All techniques are based around trapping a sample of water and some view the sediment flocs in the sample with a high resolution video camera. The techniques of University of Plymouth, HR Wallingford and Rijkswaterstaat use cameras to examine floc movements. Others use a variety of settling tubes and procedures for sampling fractions of the suspension with different settling velocities.

Different approaches measure a different range of floc sizes, e.g. some will be better at determining the highest rates than others. In some of the instruments there is a danger that the largest flocs are broken up in the collection process. Results of these measurements can be sensitive to the methods used and the methods are only suitable for use by specialists. Despite the logistical difficulties associated with fieldwork, the results are more useful than laboratory measurements, where different flocculation properties make the results less applicable to field situations. There is considerable variability between the results given by different instruments so it is recommended that if comparisons between measurements are to be made, then a single instrument type is used for all measurements.

A suspension of cohesive sediment has a range of settling velocities due to the size and density distributions of the flocs. A typical settling velocity distribution and median settling velocity of a concentration of $1\,\mathrm{kg\,m^{-3}}$ of Thames suspended sediment is shown in Figure 19 (Stevenson and Burt, 1985). The median settling velocity $w_{50}$ is the settling velocity for which half the sediment by weight will settle at a higher (or lower) velocity. Typical in-situ floc settling velocities are in the range $0{\cdot}1$–$5{\cdot}0\,\mathrm{mm\,s^{-1}}$. The floc size appears to be the dominant influence on settling

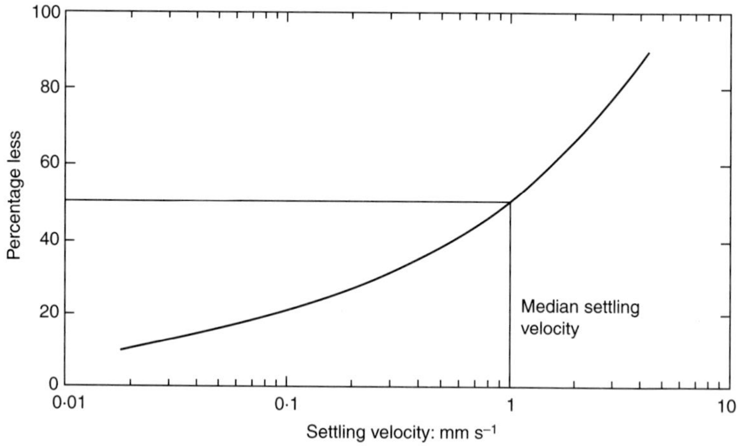


*Figure 19. Distribution of floc settling-velocity of River Thames suspended cohesive sediment against cumulative weight (Owen Tube method, after Stevenson and Burt, 1985)*

velocity of muddy sediments in estuaries. Floc size is influenced by the probability of interparticle collisions, the probability of particles sticking together when they collide and the probability of floc break-up. Increasing concentrations of suspended sediment are normally associated with more collisions and hence larger flocs and higher settling velocities, although this effect only applies when the concentration of sediment is low enough that interstitial water can escape easily. A degree of turbulent shearing increases the probability of collisions. Maximum floc size appears to be linked to the size of the smallest turbulent eddies. Vigorous turbulence leads to a reduced size of the smallest eddies, so a reduced maximum floc size (van Leussen, 1997). Maximum floc sizes in the Dollard Estuary (The Netherlands) have been found to occur one hour after maximum flood tide and at the peak of the ebb tide (Van der Lee, 1998). This implied that the flocs were not broken up by the increased turbulence for these particular conditions but that the suspended sediment concentration dominated the flocculation process.

Wherever possible it is recommended to use measurements of settling velocity obtained in-situ. A more representative description of the distribution of settling material will be obtained with one of the direct methods. Comparative results for investigating seasonal or spatial variations in settling velocity can be obtained using a gravimetric technique. Gravimetric techniques such as the Owen Tube are not able to resolve settling velocities for concentrations below about 0·1 $kg\,m^{-3}$. Hence it may be useful to use a variety of techniques to compare the settling velocity at a given site. If in-situ techniques are not available then assume a settling velocity of 1 $mm\,s^{-1}$, or use the method described below, and investigate the sensitivity of the siltation calculation to variations in this parameter.

The median settling velocity of cohesive sediment is strongly dependent on the suspended sediment concentration. Median settling velocity $w_{50}$ increases with increasing suspended sediment concentration $C_M$. The relationship may be approximated by the following empirical form of equation which is valid for dry mass concentrations up to 2–4 $kg\,m^{-3}$:

$$w_{50} = kC_M{}^m \qquad \text{SC (5.4)}$$

where $k$ and $m$ are coefficients with appropriate dimensions.

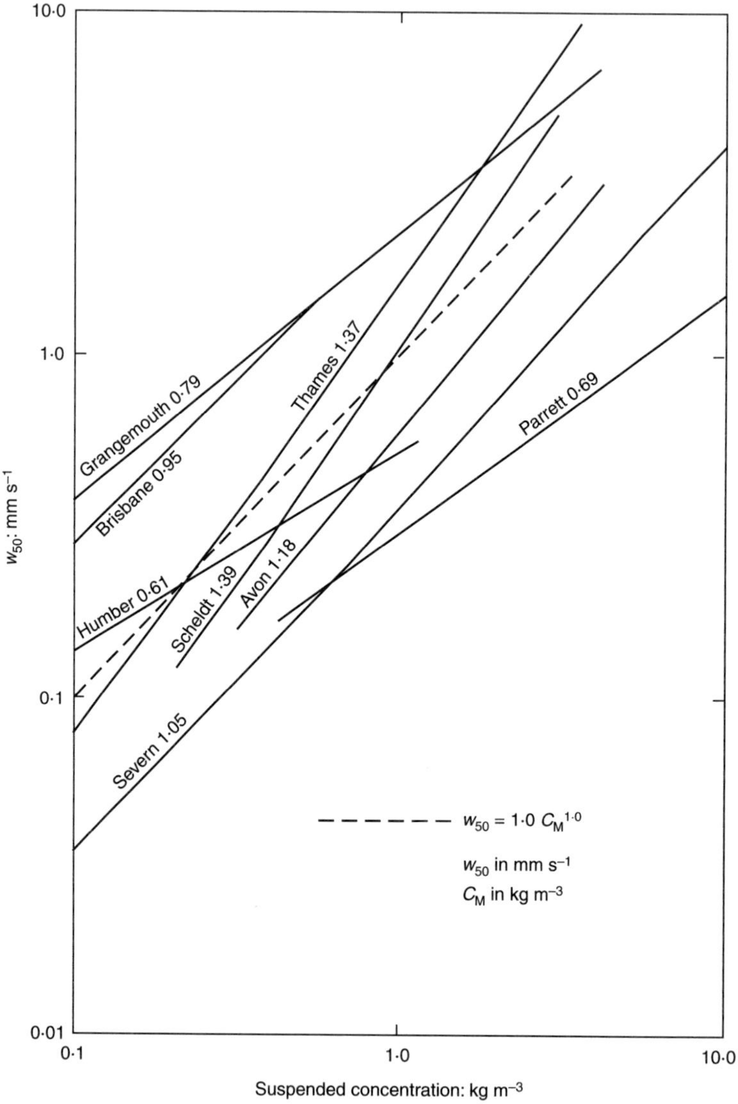


*Figure 20. Owen Tube determined median settling-velocity of suspended sediment from different sites as a function of suspended-sediment concentration $C_M$ (from Delo and Ockenden, 1992)*

The values of $k$ and $m$ vary considerably for different estuaries. A summary of results from Owen Tube experiments conducted by HR Wallingford in eight estuaries is shown graphically in Figure 20. The dashed line drawn through the middle of the data can be represented by (SI units):

$$w_{50} = 0{\cdot}001 C_{\mathrm{M}}^{1{\cdot}0} \qquad \text{SC}(5.5)$$

Thus if $w_{50}$ is in m s$^{-1}$ and $C_{\mathrm{M}}$ is in kg m$^{-3}$, then default values of the coefficients are $k = 0{\cdot}001$ and $m = 1{\cdot}0$. It can be seen in Figure 20 that there is a variability of about a factor 3 around the dashed line.

Hindered settling is the process by which a high concentration of settling flocs interfere with their surrounding flow of fluid. It usually commences at a suspended sediment concentration of between 2 kg m$^{-3}$ to 10 kg m$^{-3}$ (Krone, 1972; Burt and Stevenson, 1983; Puls and Kuehl, 1986). The settling velocity expressed as a function of suspended concentration peaks and eventually decreases at very high suspended sediment concentrations. An example of settling velocity data for the Severn Estuary is given in Figure 21.

The effect of hindered settling is sometimes expressed, by analogy with settling of very dense sand suspensions, by relationships such as:

$$w_{50} \sim (1 - C)^5 \qquad (5.6)$$

where $C$ is the volume concentration of mud.

However, an approach which covers both low and high concentrations has been developed by Soulsby (2000) from the Soulsby (1997, Equation 103) formula for the settling velocity $w_{\mathrm{sC}}$ of both low and high volume concentration $C$ of sand:

$$w_{\mathrm{sC}} = \frac{\nu}{d}\{[10{\cdot}36^2 + 1{\cdot}049(1 - C)^{4{\cdot}7} D_*^3]^{1/2} - 10{\cdot}36\} \qquad \text{SC}(5.7)$$

where $\nu$ = kinematic viscosity of water
$d$ = grain diameter
$D_*$ = a dimensionless grain diameter

This can be converted into an expression for settling of mud flocs, by assuming:

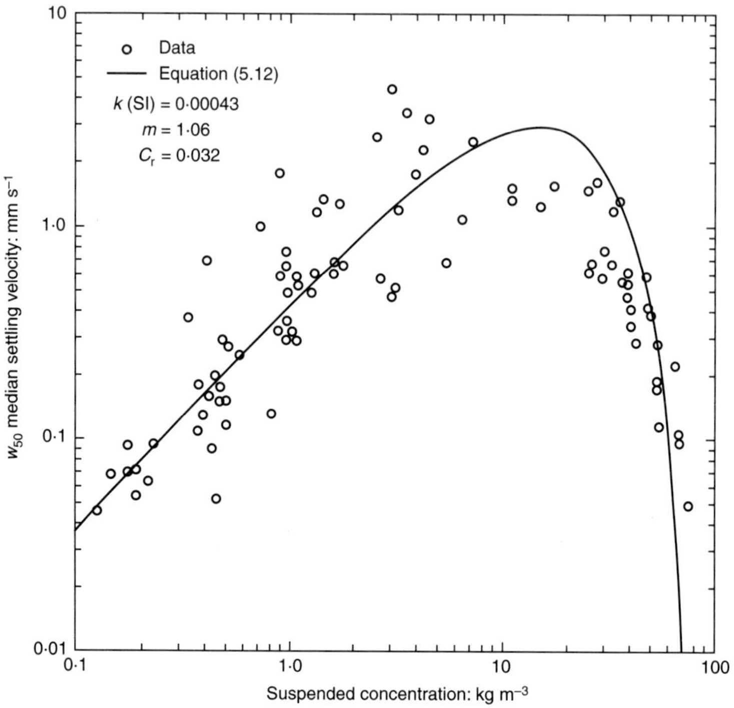


*Figure 21. Median settling-velocity of Severn Estuary mud as a function of suspended-sediment concentration (Soulsby, 2000). Equation (5.12) plotted for comparison. Owen Tube data from Odd and Rodger (1986)*

(*a*) that flocs can be treated as low-density grains (i.e. aggregation of flocs, break-up of flocs, and water-flow within flocs are all neglected)

(*b*) that the effective diameter $d_e$ of a floc increases with the volume concentration $C$ of the suspension according to the relationship

$$d_e = lC^{\frac{m}{2}} \qquad \text{SC (5.8)}$$

where $l$ is a length-scale and $m$ is the power coefficient in Equation (5.4).

(*c*) the length-scale $l$ is given by the relationship

$$l = \left[\frac{19{\cdot}8\rho\nu\rho_s^m k}{g(\rho_e - \rho)}\right]^{1/2} \qquad \text{SC (5.9)}$$

where $\rho_s$ = mineral density of the grains
$k$ = multiplicative coefficient in Equation (5.4)
$\rho_e$ = effective density of the floc

(*d*) $\rho_e = \rho + C_{in}(\rho_s - \rho)$ SC (5.10)

where $C_{in}$ is the internal volume concentration of grains inside a floc

(*e*) the volume concentration of flocs in water, $C_f$, is determined from

$$C_f = \frac{(\rho_s - \rho)C}{(\rho_e - \rho)} \quad \text{SC (5.11)}$$

It is necessary to distinguish the water-density $\rho$, the sediment grain density $\rho_s$, the effective density of the floc $\rho_e$, the volume concentration of grains in the suspension $C$, the volume concentration of sediment grains inside the floc $C_{in}$, the volume concentration of flocs in water $C_f$, and the mass (or dry) concentration of the suspension $C_M = \rho_s C$. These are all inter-related.

Making use of assumption (*a*), Equation (5.7) becomes:

$$w_{50} = \frac{\nu}{d_e}\{[10{\cdot}36^2 + 1{\cdot}049(1 - C_f)^{4{\cdot}7} D_*^3]^{1/2} - 10{\cdot}36\}$$

SC (5.12)

$$\text{where } D_* = d_e \left[\frac{g(\rho_e - \rho)}{\rho\nu^2}\right]^{1/3} \quad \text{SC (5.13)}$$

is the dimensionless floc diameter.

Equation (5.12) is equivalent to Equation (5.4) at low concentrations, but at higher concentrations a maximum settling velocity is found beyond which $w_{50}$ rapidly decreases with $C_M$. The coefficients $k$, $m$ and $C_{in}$ can be obtained by calibration against measured settling velocity curves. Variations of $k$ and $m$ are displayed in Figure 20, and values of $C_{in}$, the internal floc concentrations, are found to range from about 0·025 to 0·04, with $C_{in}$ = 0·03 as a default value. These values compare with direct estimates of $C_{in}$ from settling velocity data which show a wider range of values between 0·006 and 0·123 (Fennessy *et al.*, 1994). Comparisons of Equation (5.12) against data from the Severn Estuary, using optimised values of the coefficients $k = 0{\cdot}00043$ (SI units), $m = 1{\cdot}06$, $C_f = 0{\cdot}032$ are shown in Figure 21.

A more sophisticated treatment of the prediction of settling velocity at low and high concentrations was recently presented by Winterwerp (1999), but is too complicated to summarise here.

## *Procedure*

The settling velocity of a cohesive sediment for use in engineering studies can be obtained through one of the following approaches, depending upon the input data available and the level of detail required in the prediction.

### *New field measurements*

1. Measure settling velocity distribution in the field using an in-situ device for preferably at least six samples with a range of suspended sediment concentrations. It is now recognised that field settling tubes disrupt the largest flocs in the collection process and so may underestimate the settling velocity. Video systems give more accurate results.

   In addition, either:

2. Use video image analysis to relate the settling velocity distribution to the floc size distribution and total concentration in suspension.

   Or:

3. For each sample, plot settling velocity against percentage by weight of sample exceeding that settling velocity and determine the median settling velocity $w_{50}$ (similar to Figure 19). Calculate the arithmetic mean value of $w_{50}$ and its range.

4. If a more detailed model is required the following approach can be used. Plot the median settling velocity $w_{50}$ of each sample against suspended sediment concentration $C_M$ on log scales and fit a straight line if appropriate (similar to a single line in Figure 20). Determine the values of $k$ and $m$ in Equation (5.4). Use this equation, provided that $C_M < 3\,\text{kg}\,\text{m}^{-3}$.

5. If concentrations $C_M > 3\,\text{kg}\,\text{m}^{-3}$ are also of interest, extend the measurements of settling velocity to high concentrations. If extensive data are available then a settling velocity function can

be prescribed. For the low concentrations, determine the values of $k$ and $m$ as described above. Then try different values of $C_{in}$ in Equation (5.12) until a good fit is obtained to the maximum and the decreasing arm of the plot of $\log w_{50}$ against $\log C_m$. Then use the calibrated Equation (5.12) for predicting $w_{50}$. Equation (5.12) is available in SandCalc version 2.00.

*Existing field measurements*

6. If the site of interest has been investigated before and the data is still appropriate use data in Figure 20 or Table 8 to provide an indication of the settling velocity and tune $w_{50}$ against observed patterns of deposition.

*No field measurements*

7. If no data exists assume a settling velocity of $1\,\mathrm{mm\,s^{-1}}$ and investigate the sensitivity of the calculated siltation to variations in this parameter.

A more detailed approach can be taken as follows if required:

8. For cases in which $C_M < 3\,\mathrm{kg\,m^{-3}}$, use Equation (5.5) (i.e. $k = 0{\cdot}001$ and $m = 1{\cdot}0$) with caution as there is a significant variation about this average relationship. In a tidal estuary, settling velocities are likely to be higher during decelerating phases of the tide and around slack water, than at other times.
9. If concentrations $C_M > 3\,\mathrm{kg\,m^{-3}}$ occur, then use Equation (5.12) with values of $k = 0{\cdot}001$, $m = 1{\cdot}0$, and $C_{in} = 0{\cdot}03$. These are the default values used in SandCalc version 2.00.

*Example 5.1. Settling velocity of flocculated mud*

To calculate the settling velocity of flocculated mud in saline water, obtain estimates of the following parameters:

| | | |
|---|---|---|
| water temperature | | 10°C |
| salinity | | 35 ppt |
| density of sediment grains | $\rho_s$ | $2650\,\mathrm{kg\,m^{-3}}$ |
| mass concentration of suspended mud | $C_M$ | $10\,\mathrm{kg\,m^{-3}}$ |

Calculate the density and kinematic viscosity of the water from Figures 7 and 8 or SandCalc.

| | | |
|---|---|---|
| water density | $\rho$ | $1027\,\mathrm{kg\,m^{-3}}$ |
| kinematic viscosity | $\nu$ | $1{\cdot}36 \times 10^{-6}\,\mathrm{m^2\,s^{-1}}$ |

Use site-specific values of the coefficients $k$, $m$ and $C_{in}$ if available. In this example, use the default values $k = 0{\cdot}001$, $m = 1{\cdot}0$, $C_{in} = 0{\cdot}03$ in SI units. Then, because $C_m > 3\,\mathrm{kg\,m^{-3}}$, use Equation (5.12) to calculate $w_{50}$, or use SandCalc–Mud–Suspension–Settling Velocity–All Concentrations. If SandCalc is not available, then use Equations (5.10), (5.11), (5.9), (5.8) and (5.3) in sequence to calculate:

| | | |
|---|---|---|
| volume concentration of grains, $= C_M/\rho_s$ | $C$ | $3{\cdot}77 \times 10^{-3}$ |
| effective density of flocs | $\rho_e$ | $1076\,\mathrm{kg\,m^{-3}}$ |
| volume concentration of flocs | $C_f$ | $0{\cdot}126$ |
| length-scale | $l$ | $0{\cdot}0124\,\mathrm{m}$ |
| effective floc diameter | $d_e$ | $7{\cdot}61 \times 10^{-4}\,\mathrm{m}$ |
| dimensionless floc diameter | $D_*$ | $4{\cdot}80$ |

Then use Equation (5.12) to calculate:

| | | |
|---|---|---|
| median settling velocity | $w_{50}$ | $4{\cdot}7 \times 10^{-3}\,\mathrm{m\,s^{-1}}$ |

Note that if the low-concentration form Eq. (5.4) had been used, the value of $w_{50}$ would (incorrectly) be given as $10 \times 10^{-3}\mathrm{m\,s^{-1}}$. For low concentrations ($C_M < 3\,\mathrm{kg\,m^{-3}}$), it is valid to use Equation (5.12) or SandCalc–Mud–Suspension–Settling Velocity–Low Concentration. For example, if $C_M = 1\,\mathrm{kg\,m^{-3}}$ then Equation (5.4) gives $w_{50} = 1{\cdot}00\,\mathrm{mm\,s^{-1}}$ compared with Equation (5.12) which gives $w_{50} = 0{\cdot}937\,\mathrm{mm\,s^{-1}}$, which is very similar.

## 5.4. CONCENTRATION PROFILE

### *Knowledge*

The processes determining the variation with height above the bed of the concentration of suspended sediment (the *concentration profile*) are reasonably well understood for non-cohesive sands (Soulsby, 1997). The bed shear-stresses due to currents and/or waves act on individual grains and pick them up from the bed. Vertical water motions due to turbulence carry them higher into the water column in a diffusion-like process. The grains also experience the force of gravity which, in still water, would carry them downwards with a constant (for a given grain-size) settling velocity. The balance between the upward diffusion and the downward settling of grains gives rise to an equilibrium concentration profile, which under constant current or wave conditions can be maintained indefinitely. If the current or wave conditions vary in time or space, the concentration profile adapts reasonably quickly to the new local conditions.

However, the processes for cohesive muds appear to be different. Erosion of the bed takes place continuously if the current or wave flow conditions produce bed shear-stresses greater than the threshold of erosion, $\tau_e$. Thus the concentration of mud in suspension increases until a depth of erosion is reached at which $\tau_e$ equals the imposed shear-stress (see Figure 15). Only if the shear-stress drops below the threshold of deposition, $\tau_d$, typically about half the value of $\tau_e$, does deposition take place. The settling velocity for suspended mud is a function of the concentration, rather than the grain-size, and hence varies with height. Thus equilibrium profiles do not occur in the way they do for sands. Because settling velocities are generally much smaller for muds than sands, the concentration is more uniform through the water depth. A further consequence is that mud suspensions react very much more slowly to changes of the flow in space and time, and therefore mud put into suspension in one part of an estuary is easily carried by the currents to a distant location before it is deposited, i.e. the mud transport is dominated by advection, not by local processes. Similarly, the temporal reaction to tidal variations in current, or storm wave events, is much slower, to the extent that the 14-day spring-neap variation in tidal currents can cause greater variations in suspended mud concentration than that within the 12·5 hour tidal cycle.

Despite the above provisos, it is productive to investigate the diffusional approach to determining the concentration profile of mud in a manner analogous to that used for sand, but modified to account for the different behaviour of muds. Such an approach was explored by O'Connor and Tuxford (1980). They tried a variety of different expressions for the variation of diffusivity, settling velocity and erosion/deposition rate, resulting in a number of expressions for the concentration profile, but they did not conclude that any one model gave best predictions.

A simplified diffusional concentration profile for mud suspended by a steady current was derived by Soulsby (2000) by making the following assumptions:

(*a*) at every height $z$ the upward diffusive mass flux $F_{up}$ of mud is equal to the downward settling mass flux $F_{down}$ (i.e. the profile is in equilibrium)

(*b*) $F_{up} = -K_s \dfrac{dC_m}{dz}$

where $K_s$ is the eddy diffusivity, and $C_M$ is the dry mass concentration of suspended mud at height $z$

(*c*) $F_{down} = w_{50} C_M$, where $w_{50}$ is the median settling velocity of the mud

(*d*) the eddy diffusivity $K_s$ is constant with height and given by $K_s = 0{\cdot}0025\bar{U}h$, where $\bar{U}$ is current speed and $h$ is water depth

(*e*) the median settling velocity, for concentration $C < 3\,\mathrm{kg\,m^{-3}}$, is assumed to be given by $w_{50} = kC_M{}^m$

(*f*) the dry mass concentration of suspended mud immediately above the bed is $C_b$.

The above equations can be combined and integrated vertically to yield a diffusional concentration profile (Soulsby, 2000):

$$\frac{C_M}{C_b} = \left[1 + B\left(\frac{z}{h}\right)\right]^{-\frac{1}{m}} \quad \text{for } C_b < 3\,\mathrm{kg\,m^{-3}} \qquad \text{SC}\,(5.14)$$

$$\text{with } B = \frac{m w_{50b}}{0{\cdot}0025\bar{U}} \qquad \text{SC}\,(5.15)$$

and median settling velocity immediately above the bed

$$w_{50b} = kC_b{}^m \tag{5.16}$$

Reasonable agreement has been found between Equation (5.14) and the average of many profiles of suspended mud concentration measured in the Severn Estuary (Figure 22). (Note that individually calibrated values of $B$ were used, in preference to Equation 5.15.) However, individual measured profiles are extremely variable (Figure 23), sometimes showing smooth profiles, sometimes stepped profiles, sometimes near-bed fluid mud layers. It is likely that no single formula will be able to reproduce such variety of behaviour.

Accordingly, a simple alternative is to assume that the concentration varies linearly with height between the water surface and the bed. This leads to an empirical concentration profile (Soulsby, 2000):

$$\frac{C_M}{C_b} = 1 - \frac{(R_c - 1)}{R_c}\left(\frac{z}{h}\right) \qquad \text{SC (5.17)}$$

where $R_c$ is the ratio of the near-bed concentration to the concentration near the water surface. This ratio can be calibrated against site-specific data, or take a default value of $R_c = 3$ if no data are available.

## *Procedure*

To calculate the concentration of suspended mud at height $z$, if the concentration at the bed is known, then one of two methods may be used.

1. If a physics-based method is preferred (albeit based partly on sand rather than mud principles), then use Equation (5.14).

   The settling velocity coefficients $k$ and $m$ need to be measured using site-specific mud by the methods described in Section 5.3. If no mud is available, use default values $k = 0{\cdot}001$ (SI units), $m = 1$.

   If measured site-specific concentration profiles are available, calibrate the value of $B$ directly in Equation (5.14) from the data.

Alternatively, if no site data are available, estimate the value of $B$ as follows. In a tidal flow, use the peak current speed as the value for $\bar{U}$ at all times in the tidal cycle, since the mud suspension process responds slowly compared to the tidal period.

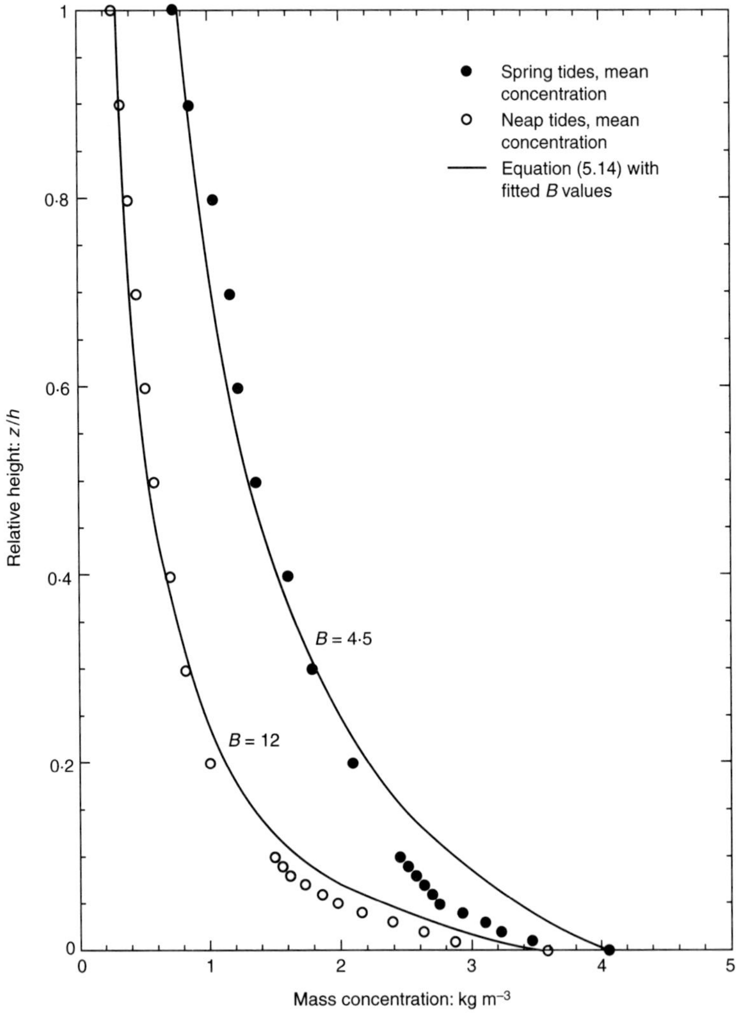


*Figure 22. Concentration profiles in Severn Estuary. Comparison of observed mean spring and neap profiles with fitted Equation (5.14) (Soulsby, 2000). Data from Kirby (1986)*

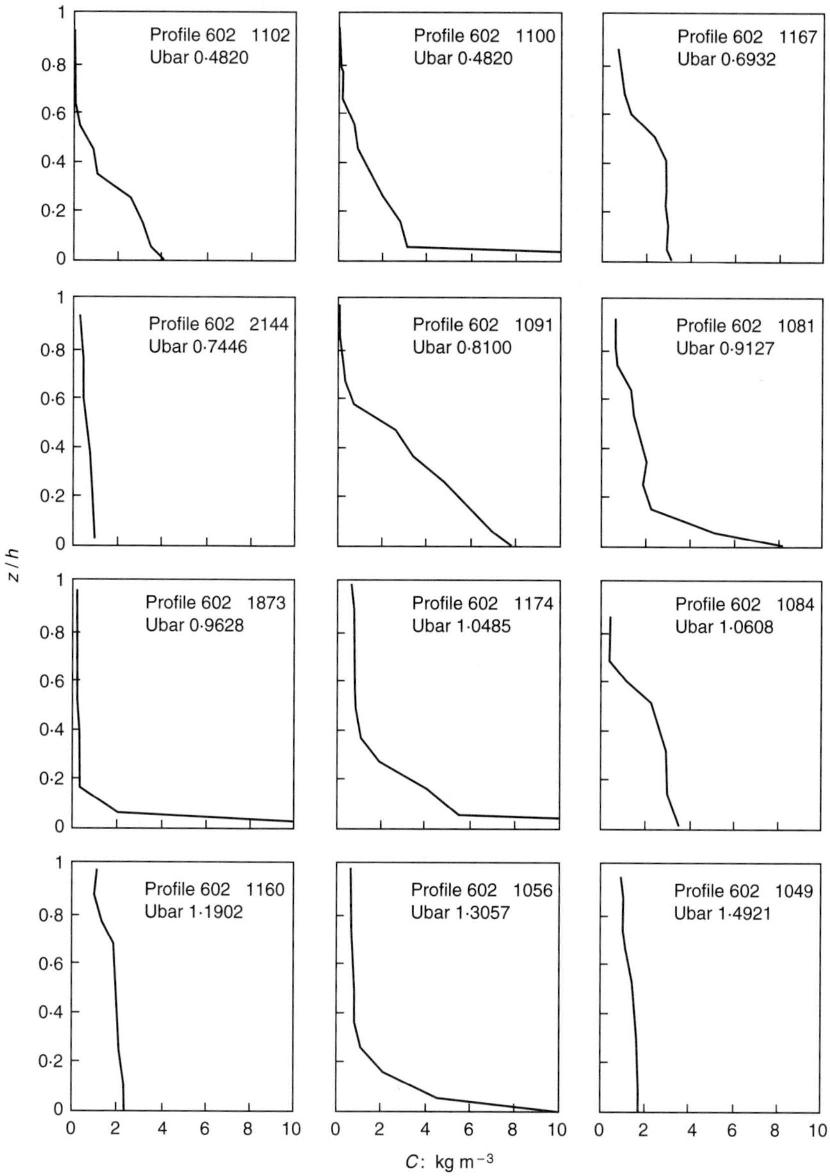


*Figure 23. Concentration profiles of mud versus relative height for the Severn Estuary. Ubar indicates current speed (in m $s^{-1}$). Data from British Oceanographic Data Centre, Severn Estuary Suspended Sediments database*

Calculate $w_{50b}$ (Equation 5.4) and then $B$ (Equation 5.15) from the above values.

Use Equation (5.14) to calculate $C_M$ at $z$.

2. If an empirical approach is preferred, calculate $R_c$ from site-specific measurements of bottom and surface concentration. If no data, use $R_c = 3$.

   Use Equation (5.17) to calculate $C_M$ at $z$.

3. In either case, be aware that the calculated concentration profiles are at best an approximation to an average profile, over time and the site, and individual profiles may differ markedly from it and, in particular, within the high concentration near-bed layers.

*Example 5.2. Suspended sediment concentration profile*
To calculate the concentration of suspended mud at a specified height above the bed suspended by a steady current, for water at 10°C, 35 ppt, $\rho = 1027\,\mathrm{kg\,m^{-3}}$, $\nu = 1{\cdot}36 \times 10^{-6}\,\mathrm{m^2\,s^{-1}}$, obtain estimates of the following parameters:

| | | |
|---|---|---|
| water depth | $h$ | 5 m |
| height at which concentration is required | $z$ | 4 m |
| concentration of suspended mud near bed | $C_b$ | $1\,\mathrm{kg\,m^{-3}}$ |
| median settling velocity near bed | $w_{50b}$ | $0{\cdot}00061\,\mathrm{m\,s^{-1}}$ |
| exponent in settling velocity formula | $m$ | 1 |
| depth-averaged current speed | $\bar{U}$ | $0{\cdot}5\,\mathrm{m\,s^{-1}}$ |

First, use the diffusional approach.

| | | |
|---|---|---|
| Calculate coefficient $B$ from Equation (5.15) | $B$ | 0·488 |
| Calculate concentration at $z = 4\,\mathrm{m}$ from Equation (5.14) | $C_M$ | $0{\cdot}610\,\mathrm{kg\,m^{-3}}$ |

Alternatively, if less information is available, use the empirical approach. Use a site-specific value of coefficient $R_c$ if available. In this example use default value $R_c = 3$.

| | | |
|---|---|---|
| Calculate concentration at $z = 4$ m from Equation (5.17) | $C_M$ | $0{\cdot}467\,\mathrm{kg\,m^{-3}}$ |

Both methods are available in SandCalc under Mud–Suspension–Concentration Profile.

# Fluid mud

# 6. Fluid mud

## 6.1. GENERAL

Although understanding of the behaviour of fluid mud is improving, there are few reliable and well-developed methods for predicting its occurrence and movement. Because of the influence of mineral types, organic material and the flocculation properties on the density and rheology of the fluid mud, its behaviour can vary considerably from one case to another. Therefore, wherever possible it is recommended to make field measurements when the influence of fluid mud is expected to be important. These must be made with care, because many field instruments are difficult to deploy or less reliable when applied very close to the bed. The material presented in this chapter represents typical fluid mud behaviour based on field and laboratory investigations and so provides a general guide for assessing fluid mud problems. However, the reader is advised to bear in mind the potential variability of fluid mud behaviour and to make measurements wherever possible. Research is underway in the scientific and engineering community to improve understanding in this area.

Fluid mud can be formed by settlement from a mud suspension in either still or flowing water (Figure 3), or by disturbance of a settled bed by wave action (see Section 6.3) or mechanical agitation. Fluid mud with suspended sediment concentrations of 70 kg $m^{-3}$ to 100 kg $m^{-3}$ have been reported in the Rivers Parrett (HR Wallingford, 1991), Severn (Kirby and Parker, 1973), Weser (Wellershaus, 1981), Gironde (Barbier, 1977), Thames (Odd and Owen, 1972) and other places around the world. Once the fluid mud layer is formed it can be moved by a gravity slope or by currents (water surface slope), or it can be mixed into the overlying water column by entrainment. The suspension becomes a fluid mud when the concentration is high enough to change the rheological properties compared with clear water. Once the sediment has become partially consolidated a significant effective stress exists in the mud layer, it ceases to be fluid mud and becomes a weak soil.

## 6.2. FLUID MUD

### *Knowledge*

There are a variety of rheological models of fluid mud, but many researchers assume a yield stress and constant viscosity (Bingham model, Figure 5 – see Section 2.6.2 for discussion). Typically, the viscosity of fluid mud is at least several hundred times larger than that of clear water and fluid mud usually exhibits shear-thinning at low rates of strain, therefore the presence of waves can reduce viscosity and increase transport rates by applying oscillating shear. Measurements in the River Parrett (HR Wallingford, 1991) of the depth and velocity of thick fluid mud layers formed by rapid settling at slack water produced no evidence of a yield stress. Velocity profile measurements could be represented by assuming a smooth turbulent flow of the fluid mud, with a thick laminar sub-layer due to the high viscosity. The best fit value for the dynamic viscosity $\mu$ was $0{\cdot}7\,\mathrm{Ns\,m^{-2}}$ compared with a value of order $0{\cdot}001\,\mathrm{Ns\,m^{-2}}$ for sea water. The condition for laminar flow in the mud layer, derived from measurements in the River Parrett was $Re = u_{\mathrm{m}}d_{\mathrm{m}}/\nu$ less than about 45, where $u_{\mathrm{m}}$ is the fluid mud velocity and $d_{\mathrm{m}}$ is the depth of the fluid mud layer with $\nu$ being the effective kinematic viscosity of the mud-water mixture. When calculating the Froude number $U/\sqrt{g'h}$ for fluid mud flow it is appropriate to use a reduced gravity term due to buoyancy, i.e. $g' = g(\rho_{\mathrm{B}} - \rho)/\rho$. This means that the speed of the fluid mud flow does not need to be very fast before it becomes supercritical, i.e. greater than 1. Fluid mud moves mainly due to bed slopes and hydrostatic pressure gradients caused by the slope of the mud-water interface or water surface, with the resistance to motion arising mainly from friction with the underlying bed. The friction between the overlying flows of water and the mud layers is fairly small owing to the suppression of turbulence by negative buoyancy of the mud. Thick layers of fluid mud move more quickly than thin layers and will start to move on less steep slopes than thin layers. This is because body forces (gravity, pressure gradients) act through the whole depth of the fluid mud, whereas resistance acts only at the bed-fluid mud interface.

As movement of pore water through fluid mud is normally very slow, it is acceptable to treat the fluid mud as a single phase fluid.

The momentum equation for a layer of fluid mud (in one dimension) averaged over the fluid mud layer of thickness $d_m$ and bulk density $\rho_B$, is:

$$\rho_B \frac{\partial u_m}{\partial t} = -g\rho \frac{\partial \eta}{\partial x} - g(\rho_B - \rho)\frac{\partial \eta_m}{\partial x} + \frac{\tau_i - \tau_b}{d_m} \quad (6.1)$$

where $\eta$ = water surface level
$\eta_m$ = fluid mud-water interface level
$\tau_i$ = shear stress at fluid mud-water interface
$\tau_b$ = shear stress at fluid mud-bed interface

This assumes a constant fluid mud density and neglects non-linear and diffusion terms for simplicity because they are usually small. The expression uses a Newtonian assumption for the rheological behaviour of the fluid mud, this can be changed to a Bingham fluid by adding the yield stress into the final term (Section 2.6.2).

It is noted that because the term involving the slope of the mud-water interface is multiplied by the density difference, rather than the density, then the water surface slope is about 20–25 times more effective (for a typical fluid mud dry-density of 70 kg $m^{-3}$) than the mud surface slope in moving mud. Thus it is possible for a sloping water surface to move mud up a shallow bed slope, but if the bed slope is steep, then fluid mud will flow down it regardless of water movement. Less dense fluid mud layers are more influenced by water slope and less influenced by interface slope.

The settling velocity of fluid mud reaches a peak of approximately 2–4 mm $s^{-1}$. The flocs hinder the displacement of the water as they settle and although the flocs touch each other, they are too weak to transmit significant forces without deforming. The settling velocity decreases rapidly as the suspended sediment concentration increases above 20 kg $m^{-3}$ and may typically be 0·05 mm $s^{-1}$ at 70 kg $m^{-3}$ (see Figure 21).

In the absence of a significant amount of vertical turbulent exchange near slack water, the net flux of settling particles is the product of the settling velocity and the concentration of mud in suspension near the bed. This flux rises to a maximum of around $60 \times 10^{-3}\,\text{kg}\,\text{m}^{-2}\,\text{s}^{-1}$ at a suspended sediment concentration of approximately $25\,\text{kg}\,\text{m}^{-3}$ and reduces at higher suspended sediment concentrations as shown in Figure 24. Fluid mud with a suspended sediment concentration of $70\,\text{kg}\,\text{m}^{-3}$ deposits on to the bed typically at a rate of about $1 \times 10^{-3}\,\text{kg}\,\text{m}^{-2}\,\text{s}^{-1}$.

A layer of fluid mud above the bed will only grow in thickness if the flux of mud settling on to the bed exceeds the rate at which the fluid mud dewaters to become a weak soil. This is equivalent to having a suspended sediment concentration of approximately $2{\cdot}5\,\text{kg}\,\text{m}^{-3}$ in the overlying water prior to slack water.

If the shear stress between a moving fluid mud layer and the bed falls below $0{\cdot}1\,\text{N}\,\text{m}^{-2}$ it will dewater. Fluid mud dewaters at a rate of approximately $0{\cdot}05\,\text{mm}\,\text{s}^{-1}$ to form a weak soil with a dry mass density of $100\,\text{kg}\,\text{m}^{-3}$ to $300\,\text{kg}\,\text{m}^{-3}$ which

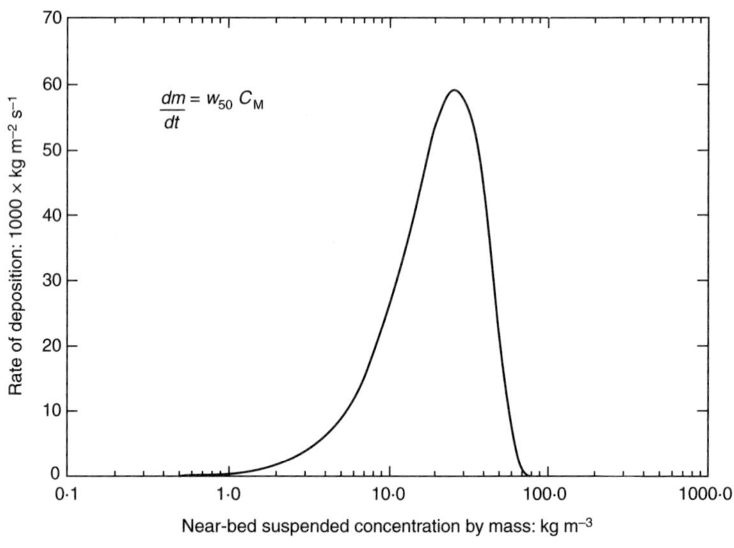


*Figure 24. Rate of deposition of suspended sediment as a function of near-bed suspended-sediment concentration (Soulsby, 2000). $w_{50}$ calculated using Equation (5.12)*

does not flow easily (Odd and Cooper, 1989; HR Wallingford, 1991).

## Procedure

The following approaches can be adopted to monitor the formation and characteristics of fluid mud at a particular site. A simple prediction method can be adopted if field data are not available, however, whenever possible predictions should be checked against appropriate field data.

### *Field and laboratory measurements*

1. Measure suspended sediment concentrations by sampling or optical techniques. If the suspended sediment concentration is greater than $2{\cdot}5\,\text{kg}\,\text{m}^{-3}$ preceding slack water, fluid mud is likely to be formed. A density measuring probe (e.g. piezo-electric or gamma ray) can be used to obtain values of density in the fluid mud layer.
2. Measure dewatering rates in the field using echo sounding techniques to record the fluid mud-water interface.

   Or:
3. Conduct laboratory tests on fluid mud pumped out of the field to determine dewatering rates.

### *No field measurements*

4. Calculate the flux of settling mud to the bed according to Figure 24.
5. Assume a constant dewatering rate of $0{\cdot}05\,\text{mm}\,\text{s}^{-1}$.

## 6.3. FLUIDISATION BY WAVES

## Knowledge

Fluidisation (or liquefaction) of the bed is caused by shear-stresses in the pore water, but without significant movement of the pore water, and an associated build up in the pore water pressure. Liquefaction tends to happen in sediments of low permeability, in relatively unconsolidated conditions, while waves acting on over-consolidated sediments are more likely to produce erosion of surface particles. Thus an important question to address is when will waves cause fluidisation and when will they erode surface particles?

Under cyclic loading by waves, the structure of the bed may be progressively weakened. The mud bed responds in both an elastic and a viscous manner; the elastic response is in the form of a restoring force (restoring the bed to its undisturbed position) while the viscous response is in the form of a dissipative force. Eventually, there is a complete breakdown of the structure (fluidisation), which allows the mud bed to be eroded or entrained much more easily. Fluidisation of the mud is dependent on the wave characteristics and the mud properties (Derbyshire and Kendrick, 1987). Transport of mud on a horizontal bed can occur owing to the mass transport velocity under waves.

The damping of waves over a fluid mud bed, showing the dissipation of wave energy, has been observed in flumes and in the laboratory. Hence, the estimation of shear-stress at the bed surface for the purpose of correlation with the rate of erosion should not necessarily be based on the assumption of a rigid bed (Maa and Mehta, 1985).

## *Procedure*

The formation conditions for fluid mud to be produced by waves and/or currents can be assessed as follows:

### *Laboratory measurements*

1. Conduct laboratory tests under combined waves and currents to determine a critical wave shear-stress for fluid mud generation and movement or entrainment into suspension.

### *No field or laboratory measurements*

2. Calculate the maximum combined wave and current shear-stress from the current velocities and wave conditions use Equation (3.26).

3. Calculate the mass which is fluidised, $M$, by assuming that the bed is fluidised to a depth (and density) at which the shear strength of the bed (Figures 15, and 13 or 14) is equal to the

combined wave and current shear-stress. Assume that the fluidised material is contained within a thin layer close to the bed. Calculate the thickness of the layer, $d_m$, according to:

$$d_m = \max(M/C_0, \delta) \tag{6.2}$$

where $d_m$ = thickness of the fluidised layer (m)
$M$ = dry mass per unit area which is fluidised ($kg\,m^{-2}$)
$C_0$ = dry mass concentration in fluid mud layer, typically $75\,kg\,m^{-3}$
$\delta$ = wave boundary layer thickness
$= (f_w/2)^{0.5} U_w T/4\pi$
$f_w$ = wave friction factor
$U_w$ = bottom orbital velocity ($ms^{-1}$)
$T$ = wave period (s)

The wave friction factor and bottom orbital velocity can be calculated according to *Procedure* in Section 3.3. Roberts (1993) developed a fluid mud model which included the influence of the mud viscosity on the boundary layer thickness.

## 6.4. MOVEMENT OF FLUIDISED LAYER

### *Knowledge*

Once the mud has been fluidised it will move more easily than a structured bed. It may flow under forces due to a sloping bed, the water surface slope or the mud-water interface slope. It may also be entrained into the overlying water.

The effects of the bed slope or the water surface slope are generally much more significant than the effect of the interface slope between the fluid mud and the overlying water, unless the interface slope is the only forcing factor (e.g. open water dumping of mud over a flat bed).

Field observations (HR Wallingford, 1991) have not shown any indication of a significant shear-stress below which the fluid mud did not move. However, there was a strong tendency for the fluid mud to dewater and consolidate with low bed stresses in the order of 0·1 $Nm^{-2}$. Measured velocity profiles in the fluid mud layer could be predicted by application of a smooth turbulent theory, allowing a thick laminar sub-layer (in which viscous forces predominate) and a smooth turbulent region with a logarithmic velocity profile.

For fluid mud moving down a bed slope, the ratio of the shear stress at the interface between the water and mud, $\tau_i$, to the shear stress at the bed, $\tau_b$, is approximately 0·43 (Harleman, 1961).

For fluid mud moving under a water surface slope, the ratio of the shear stress at the interface, $\tau_i$, to the shear-stress at the bed, $\tau_b$, is a function of the water depth:

$$\tau_i = \tau_b(1 - d_m/h) \qquad (6.3)$$

where $\tau_b$ = shear stress at bottom of fluid mud
$\tau_i$ = shear stress at interface of mud and water
$d_m$ = depth of fluid mud
$h$ = depth of overlying water

## Procedure

The sediment transport rate of fluid mud can be estimated as follows.

### Field measurements

1. Use field equipment to measure the velocity profile in the fluid mud layer and in the overlying water column during formation and movement of the fluid mud layer. Combined with the measurement of the bed-density profile, or average density of the fluid mud layer, the product of the velocity and concentration produces an estimate of the flux of mud.

### No field measurements

2. If the mud is on a slope, use the depth of the fluid mud layer to calculate the velocity $u_m$ of the layer due to the bed slope from Figure 25 (after Odd and Rodger, 1986; Delo and Ockenden, 1992). Figure 25 has been calculated assuming that smooth turbulent theory applies in the fluid mud layer. It assumes a uniform dry density for the fluid mud layer of 1075 $kg\,m^{-3}$, and a uniform dynamic viscosity of 0·7 $Ns\,m^{-2}$. This has been found

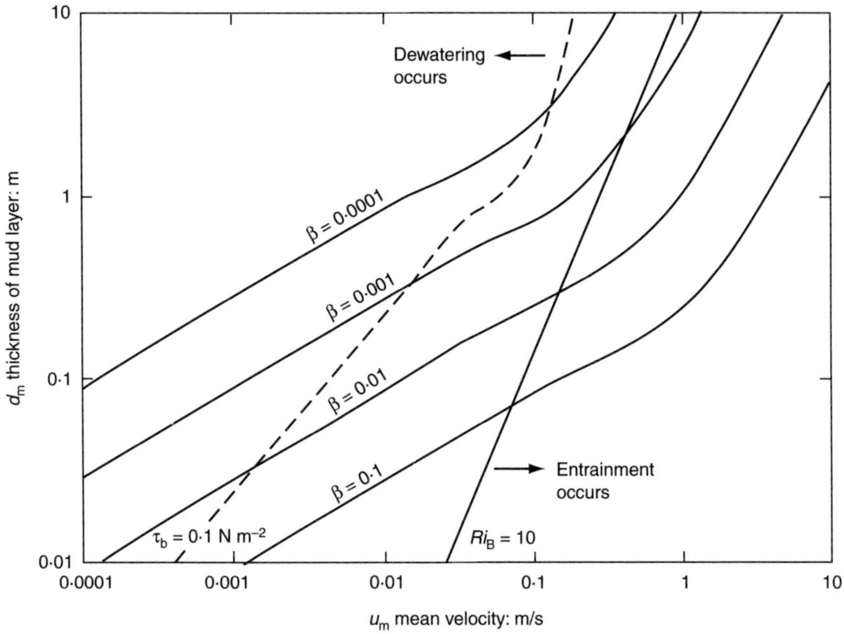


*Figure 25. Mean velocity $u_m$ of a layer of fluid mud of thickness $d_m$ down a sloping bed in still water, for a range of values of bedslope $\beta$, after Odd and Rodger (1986) (from Delo and Ockenden, 1992). Limits for dewatering and entrainment shown*

to be a reasonable assumption (HR Wallingford, 1991). The value of $u_{\mathrm{m}}$ is obtained knowing $d_{\mathrm{m}}$ and the value of $\beta$.

3. If the layer is subject to a water surface slope, calculate the water surface slope and calculate the velocity of the layer moving due to this water surface slope from Figure 26 (Delo and Ockenden, 1992). The value of $u_{\mathrm{m}}$ is obtained knowing $d_{\mathrm{m}}$ and the value of $\partial\eta/\partial x$.

4. Calculate the sediment transport rate, $Q_{\mathrm{m}}$, according to:

$$Q_{\mathrm{m}} = \rho_{\mathrm{m}} d_{\mathrm{m}} u_{\mathrm{m}} \tag{6.4}$$

Typically, the fluid mud velocity can be about half the depth-averaged current speed, where the transport is due to the water surface slope.

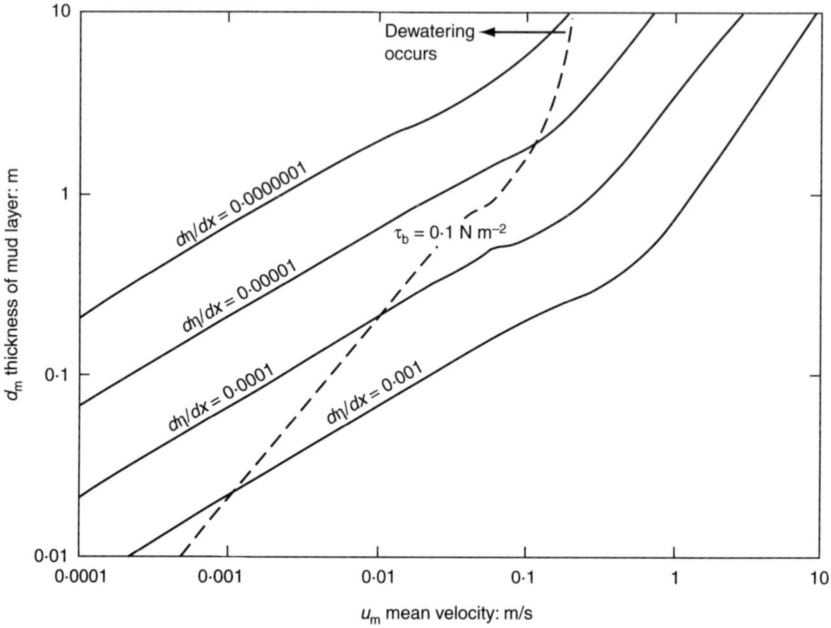


*Figure 26. Mean velocity $u_m$ of a layer of fluid mud of thickness $d_m$ under an applied water surface slope, for a range of values of water surface slope $d\eta/dx$ (from Delo and Ockenden, 1992). Limit for dewatering shown*

## 6.5. ENTRAINMENT OF FLUIDISED LAYER

### *Knowledge*

The entrainment of clear water into a fluid mud layer appears to behave in a similar way to the mixing of a salt wedge. Entrainment occurs if the bulk Richardson number, $Ri_b$, is lower than about 10 (HR Wallingford, 1990, 1991). $Ri_b$ is a discretised version of the gradient Richardson number, see Equation (5.1).

The bulk Richardson number $Ri_b$ is:

$$Ri_b = \frac{g d_m \Delta\rho}{\rho(\Delta u)^2} \tag{6.5}$$

where $\Delta\rho$ = density difference between fluid mud and water, i.e. $\rho_B - \rho$ in which $\rho$ includes the contribution to the overlying fluid density from suspended sediment mass

$\Delta u$ = difference between the fluid mud velocity and the overlying water velocity

The entrainment velocity, i.e. the rate at which the thickness of the fluid mud layer is reduced as material is entrained from its upper surface, varies as a function of $Ri_b$. Following the saline wedge analogy entrainment rate or vertical flux of mud ($kg\,m^{-2}\,s^{-1}$) may be approximated by:

$$\frac{dm}{dt} = -V_e \Delta u C_0 \quad \text{for } Ri_B < 10 \tag{6.6}$$

where $C_0$ is the constant sediment concentration of the fluid mud layer and the entrainment coefficient $V_e$ is given by:

$$V_e = u_m \frac{0{\cdot}1}{(1 + 63 Ri_b^2)^{0{\cdot}75}} \tag{6.7}$$

However, a comparison between Equation (6.7) and field data for fluid mud (HR Wallingford, 1991) has indicated that $V_e$ could be an order of magnitude higher than is given by Equation (6.7).

## Procedure

The entrainment rate of fluid mud into the overlying water column can be estimated as follows:

1. Calculate the bulk Richardson number according to Equation (6.5). If $Ri_B < 10$, calculate the entrainment rate according to Equations (6.6) and (6.7).

   Or:

2. Use Figure 25 to estimate whether mud is entrained given the depth and velocity of the layer.

# Transport rate

# 7. *Transport rate*

## 7.1. GENERAL

Once the sediment is removed from the bed (by erosion or entrainment) it is important to calculate the rate at which the sediment is transported away by the current.

## 7.2. TRANSPORT RATE

### *Knowledge*

The transport rate of mud is defined as the dry mass of sediment crossing a unit width of bed in unit time. Thus in SI units the transport rate has the units $\mathrm{kg\,m^{-1}\,s^{-1}}$. The transport rate of suspended mud, $Q_{\mathrm{smud}}$, can be obtained from the product of the concentration profile and the velocity profile, integrated over the water depth:

$$Q_{\mathrm{smud}} = \int_0^h C_{\mathrm{M}}(z)U(z)\,dz \tag{7.1}$$

where $h$ = water depth
$C_{\mathrm{M}}(z)$ = mass concentration (dry density) of mud at height $z$
$U(z)$ = current speed at height $z$

The exact definition of what constitutes the level of the bed, $z = 0$, is not always clear-cut in settling and consolidating mud suspensions. To ensure that Equation (7.1) includes all the transport, the integration should extend down to a level at which the bed is immobile ($U = 0$), and this then defines $z = 0$.

Note that the transport rate given by Equation (7.1) is not the same as the product of depth-averaged velocity and the depth-averaged concentration.

If waves are present, then strictly speaking the transport rate at any instant in the wave cycle is given by Equation (7.1), with $C(z)$ and $U(z)$ being instantaneous concentration and velocity profiles. The net transport rate is then given by taking a time-average of Equation (7.1). This approach is used in the most recent formulations of sand transport, where the correlation between velocity and concentration can give rise to a substantial wave-related contribution to the total transport. However, the understanding of mud suspension under waves is not yet sufficiently advanced to include this refinement, and for practical purposes it is usual to apply Equation (7.1) with $C(z)$ and $U(z)$ taken as the time-averaged concentration and velocity profiles. These do take account of the net effect of wave-stirring, but not the cross-correlation effect.

If fluid mud is present, Equation (7.1) applies through the fluid mud layer. Fluid mud layers may move as a result of horizontal pressure gradients, shear-stress from overlying water, and sloping beds (see Section 6.4). The current velocity within a fluid mud layer can be quite high – typically about half of the depth-averaged current-speed where it is driven by the pressure gradient due to the tide. When this is multiplied by the very high concentrations within the fluid mud layer, the transport within the layer may be many times larger than that in the overlying, more dilute, suspension. Hence it is important to ascertain whether fluid mud is likely to be present at a study site, and to measure the concentration, thickness, extent and lifetimes of such layers.

## *Procedure*

The transport rate of mud in suspension can either be determined from measurements or, in the absence of data, predicted using a semi-theoretical approach as follows:

### *Field measurements*

1. Use field equipment to measure the concentration and current speed at a number of levels between the bed and the water

surface, and at intervals through a tidal cycle. The levels should be closely spaced near the bed and more widely spaced in the upper part of the water column. The interval between profiles should be 30 to 60 minutes, and readings should be made on mean spring tides and preferably also on mean neap tides. Measure through a fluid mud layer, if present.

2. Perform a numerical integration as a finite-difference form of Equation (7.1). Assume that the velocity at $z = 0$ is zero, and the velocity and concentration at the water surface are equal to that at the highest measured level, providing this is in the upper half of the water column. The integration is given by a trapezium rule.

   Thus, if the current speeds at heights $z_1, z_2, z_3, \ldots, z_n$ (increasing upwards) are $U_1$, $U_2$, $U_3$, ..., $U_n$, and the concentrations are $C_1, C_2, C_3, \ldots, C_n$, then the calculated value of the transport rate is:

$$\begin{aligned} Q_{smud} = 0.5[&C_1U_1z_1 + (C_1U_1 + C_2U_2)(z_2 - z_1) \\ &+ (C_2U_2 + C_3U_3)(z_3 - z_2)\ldots \\ &+ (C_{n-1}U_{n-1} + C_nU_n)(z_n - z_{n-1}) \\ &+ 2C_nU_n(h - z_n)] \end{aligned} \quad (7.2)$$

   Equation (7.2) is an adequate approximation to $Q_{smud}$ provided that: (*a*) there are sufficient measurement heights (preferably at least 6); (*b*) the lowest height is sufficiently close to the bed (not higher than 0·2 m); (*c*) the highest height is above mid-depth; and (*d*) the measurements are either simultaneous, or measured over a time less than about 20 minutes.

*No field measurements*

3. Calculate the mass of sediment in suspension, and hence the depth-averaged concentration, from the time-history of erosion and deposition as a function of bed shear-stress.

4. Calculate the theoretical concentration profile using Equation (5.14) if sufficient information is available, or Equation (5.17) if it is not. Scale the concentration profile so as to give the calculated depth-averaged concentration.

5. Calculate the tidal current velocity profile from the following equation (Soulsby, 1997):

$$U(z) = \left(\frac{z}{0{\cdot}32h}\right)^{1/7} \bar{U} \quad \text{for } 0 < z \leqslant 0{\cdot}5h \qquad \text{SC (7.3)}$$

$$U(z) = 1{\cdot}07\bar{U} \quad \text{for } 0{\cdot}5h < z < h \qquad \text{SC (7.4)}$$

   This equation is easier to use than the logarithmic profile Equation (3.3) because it does not involve the parameter $z_0$, which is difficult to evaluate. Whitehouse (1993) found that Equations (7.3) and (7.4) gave good agreement with measured velocity profiles in the outer Thames estuary. However, Equation (3.3) has a sounder theoretical basis and can alternatively be used. The difference between the two methods is generally less than 10%.

6. Perform a numerical integration of the predicted velocity and concentration profiles using Equation (7.2), with up to 50 levels chosen to be very closely spaced near the bed, and more coarsely spaced higher up.

7. If fluid mud is present, the concentration within that layer must be calculated separately (see Sections 6.2, 6.3 and 6.4). The velocity profile goes to zero at the bottom of the fluid mud layer, and as an approximation increases linearly to the top of the layer. The integration of the dilute suspension layer above this must be altered accordingly. Prediction of the presence and concentration of fluid mud is difficult (although step 1 of *Procedure* in Section 6.2 may help). If fluid mud is known to be prevalent at a study site then field measurements should be made whenever possible.

8. If layers with strong gradients in concentration (*lutoclines*) or salinity (*haloclines*) are present, then the velocity profile may be modified significantly from the simple unstratified forms. In such cases either field measurements or a more sophisticated model is required.

## 7.3. TRANSPORT RATE FOR A VISCOUS LAYER

### *Knowledge*

Even quite dense (dry mass concentration $> 200\,\text{kg m}^{-3}$) deposits of mud can move with time, with the rate of spreading much too slow to be described by the movement of fluid mud. The mud can be described theoretically as a viscous layer of thickness $d_m$ moving down a slope of angle $\beta$. For viscous flow, with constant viscosity

$$\tau(z) = \mu\, dU_m/dz \tag{7.5}$$

where $\tau(z)$ = shear stress at height $z$ in the bed
$\mu$ = dynamic viscosity of the mud ($\mathrm{N\,s\,m^{-2}}$)
$U_m(z)$ = velocity of the mud at height $z$ in the bed

## *Procedure*

The transport rate of a mud deposit on a slope can be calculated using the following method:

1. Assume a constant bulk density, $\rho_B$, and constant viscosity (reasonable for a thin layer). Calculate the force balance:

$$(\rho_B - \rho)g \sin\beta + d\tau/dz = 0 \tag{7.6}$$

2. Assume a no-slip condition at the bottom of the layer. Then the velocity profile in the layer is:

$$U(z) = (\rho_B - \rho)g \sin\beta(d_m - 0{\cdot}5z^2)/\mu \tag{7.7}$$

3. Integrate through the layer to obtain the layer-averaged velocity of the mud, $\bar{U}_m$:

$$\bar{U}_m = \frac{(\rho_B - \rho)g d_m^2 \sin\beta}{3\mu} \qquad \text{SC (7.8)}$$

4. The density of the mud has been assumed constant throughout the layer. Hence, the dry mass transport rate $Q_{bmud}$ ($\mathrm{kg\,m^{-1}\,s^{-1}}$) of mud moving down the slope is given by the product of $\bar{U}_m$, the dry density $C_m$, and the layer thickness $d_m$:

$$Q_{bmud} = \frac{C_M(\rho_B - \rho)g d_m^3 \sin\beta}{3\mu} \qquad \text{SC (7.9)}$$

5. However, if the flow velocity becomes too large, the assumption that the flow is laminar breaks down. Hence, Equation (7.9) is only valid if the layer Reynolds number is less than 45:

$$\frac{\rho_B \bar{U}_m d_m}{\mu} < 45 \tag{7.10}$$

If values of the input parameters are unknown:

6. Use a dry density in the range $C_M = 150 - 300\,\mathrm{kg\,m^{-3}}$
7. Derive bulk density $\rho_B$ using Figure 1.
8. Use a dynamic viscosity in the range $\mu = 0{\cdot}2 - 0{\cdot}7\,\mathrm{Ns\,m^{-2}}$.

*Example 7.1. Mud flow down a slope*
A partially consolidated layer of mud lies on mudflats at the edge of an estuary which have a slope of 1:100 towards the deep channel. To calculate the rate at which the mud flows down the slope into the channel due to gravity alone, obtain estimates of the following parameters:

| | | |
|---|---|---|
| bulk density of the mud layer | $\rho_B$ | $1200\,\mathrm{kg\,m^{-3}}$ |
| thickness of mud layer | $d_m$ | $0{\cdot}1\,\mathrm{m}$ |
| dynamic viscosity of mud | $\mu$ | $0{\cdot}5\,\mathrm{Ns\,m^{-2}}$ |
| density of sediment grains | $\rho_s$ | $2650\,\mathrm{kg\,m^{-3}}$ |
| density of water (10°C, 35 ppt) | $\rho$ | $1027\,\mathrm{kg\,m^{-3}}$ |
| angle of slope of bed (1:100) | $\beta$ | $0{\cdot}573°$ |

Calculate the layer-averaged velocity of the mud from Equation (7.8).

| | | |
|---|---|---|
| mud velocity | $\bar{U}_m$ | $0{\cdot}113\,\mathrm{m\,s^{-1}}$ |

Calculate the Reynolds number for the mud layer.

| | | |
|---|---|---|
| $\rho_B \bar{U}_m d_m / \mu$ | *Re* | 27·1 |

The Reynolds number is well below the critical value of 45, so the mud flow is laminar.

Calculate the dry density from the bulk density using Figure 1.

| | | |
|---|---|---|
| dry density | $C_M$ | $282\,kg\,m^{-3}$ |

Calculate the transport rate from Equation (7.9), or use Sand-Calc–Mud–Mud Flow–On Slope.

| | | |
|---|---|---|
| bedload transport rate of mud | $Q_{bmud}$ | $3{\cdot}19\,kg\,m^{-1}\,s^{-1}$ |

# Deposition

# 8. Deposition

## 8.1. GENERAL

Owing to the fact that the speed of the tidal current, water depth and wave conditions vary with time there will be periods of low flow around slack water when sediment is able to settle out of the water column and become deposited on the bed (Figure 3). The settling velocity of sediment is governed by the properties of the material in suspension and the ambient water properties as has been discussed in Section 5.3.

## 8.2. DEPOSITION IN STILL WATER

### *Knowledge*

The rate of deposition of sediment to the bed can be described by the near-bed concentration of suspended sediment $C_b$ and its median settling velocity $w_{50}$. In still water, the rate of deposition of sediment from suspension $dm/dt$ can be expressed by:

$$\frac{dm}{dt} = -C_b w_{50} \tag{8.1}$$

There is likely to be a gradient of concentration of suspended sediment in the water column with the near-bed concentration greater than the depth-mean average value (e.g. Figure 22). Therefore the concentration of suspended sediment in the water column will decrease with time as sediment deposits on the bed and the rate of deposition will decrease accordingly.

The rate of deposition of suspended sediment as a function of the near-bed dry mass suspended sediment concentration is plotted for typical parameters in Figure 24. The value of settling velocity $w_{50}$ was calculated with Equation (5.12) across the whole range of concentrations.

## Procedure

Methods for determining the rate of deposition are as follows:

*New field measurements*

1. The instantaneous rate of deposition can be calculated from the measured median settling-velocity relationship and the near-bed suspended sediment concentration with Equation (8.1).

*Existing field measurements*

2. If the site of interest has been investigated before and the data are still suitable obtain appropriate values of $w_{50}$ from Figure 20. Other site specific values are given in Table 8. The rate of deposition is then calculated using Equation (8.1).

*No field measurements*

3. Use the average settling velocity relationship given in Equation (5.12) with caution and the near-bed suspended sediment concentration to calculate the deposition rate with Equation (8.1).

# 8.3. DEPOSITION IN FLOWING WATER

## Knowledge

Laboratory tests have been conducted in straight and circular flumes to investigate the mechanism of deposition of cohesive sediment (Krone, 1962; Partheniades, 1962; Postma, 1962; Partheniades *et al.*, 1966; Partheniades *et al.*, 1968; Mehta and Partheniades, 1973; Mehta, 1988; Burt and Game, 1985; Delo 1988b). Also Owen (1971) and Wolanski *et al.* (1992) have examined the role of turbulence on the settling of mud flocs.

Field measurement of the deposition of cohesive sediments in estuaries during single tides has been undertaken by HR Wallingford (Diserens *et al.*, 1991). The hydrodynamics of the flow (including wave-induced currents), suspended sediment concentration and the bed elevation were recorded.

The rate of deposition of cohesive sediment from a suspension in flowing water can be modelled using the near-bed concentration $C_b$, median settling velocity $w_{50}$, bed shear-stress $\tau_0$ exerted by the flowing water and a critical bed shear-stress $\tau_d$ for deposition. This is defined as the bed stress above which there is no deposition of suspended sediment. For a uniform sediment, or one for which the settling velocity is approximated by the median settling velocity, an empirical equation can be used to calculate the rate of deposition:

$$dm/dt \begin{cases} = -(1 - \tau_0/\tau_d)C_b w_{50} & \text{for } \tau_0 < \tau_d \\ = 0 & \text{for } \tau_0 \geqslant \tau_d \end{cases} \qquad \text{SC (8.2)}$$

The critical bed shear-stress for deposition of cohesive sediment $\tau_d$, is estimated from laboratory tests to be between $0{\cdot}06\,\mathrm{N\,m^{-2}}$ and $0{\cdot}10\,\mathrm{N\,m^{-2}}$. It is typically about half the value of the threshold bed shear-stress for erosion, $\tau_e$, but is not directly related to $\tau_e$.

Suspended cohesive sediments can be considered to consist of flocs which have a distribution of sizes, densities, settling velocities, strengths and critical shear-stresses for deposition. The representation of these suspended sediments at a particular concentration by a median settling velocity and a single critical shear-stress for deposition of all the sediment is therefore an approximation. Part of the suspended sediment will deposit at bed shear-stresses greater than the 'average' critical stress given above. Figure 27 shows the fraction of sediment remaining in suspension following deposition in flowing water (Delo, 1988b). The deposition of a suspended cohesive sediment with a range of settling velocities has been investigated and modelled by Mehta (1988), Mehta and Lott (1987), Delo (1988b), Krishnappan (1991), Verbeek *et al.* (1991) and Ockenden (1993), but this approach requires more input information than for a uniform sediment.

A 'distributed' sediment can be modelled by dividing the sediment into $N$ classes, each with a unique settling velocity, $w_{si}$, concentration, $c_i$, and critical shear-stress for deposition,

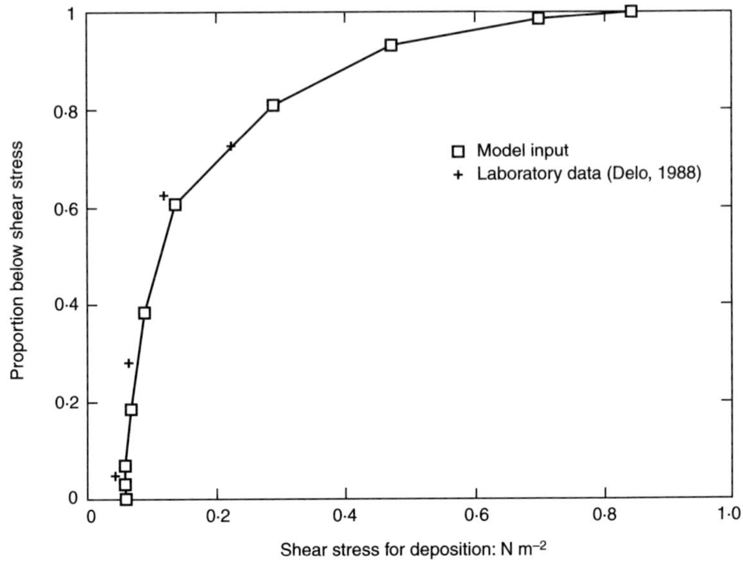


*Figure 27. Critical shear-stress for deposition of cohesive sediment from laboratory Carousel tests (data from Delo, 1988b)*

$\tau_{di}$. The total amount of sediment deposited on the bed, $\Delta m$, during a time interval, $\Delta t$, is given by the sum of the individual amounts deposited from each class (Ockenden, 1993):

$$\Delta m = \sum_{i}^{N} w_{si} \phi C_i C(t)(1 - \tau_0/\tau_{di})\Delta t \tag{8.3}$$

where $\phi C_i$ = proportion of the total concentration in sediment class $i$

$C(t)$ = suspended sediment concentration at time $t$

Sediment class $i$ only deposits if $\tau_0 < \tau_{di}$. Each class of sediment flocs is assumed to act independently of the others.

## Procedure

Methods for determining the rate of deposition in flowing water are as follows.

*New field and laboratory measurements*

1. The instantaneous rate of deposition can be calculated from the measured median settling-velocity relationship, the near-bed suspended-sediment concentration, the shear-stress at the bed and an assumed value of $\tau_d$ ($0{\cdot}08\,\mathrm{N\,m^{-2}}$) using Equation (8.2).

2. A field system can be deployed to measure the hydrodynamics, suspended-sediment concentration and bed elevation through a tide. The data can be used to determine the in-situ rate of deposition. In addition, a deposition algorithm such as that given by Equation (8.2) can be calibrated and evaluated.

3. Use video image analysis or similar equipment to make measurements of the settling velocity distribution and floc size distribution. Use this to model the deposition of distributed sediment according to Equation (8.3).

4. Determine the critical shear-stress for deposition (single value or value for each sediment band) either from laboratory deposition tests in a straight or annular flume or from field data (item 2).

*Existing field measurements*

5. If the site of interest has been investigated before and the data are still appropriate obtain an appropriate value of the settling velocity from Figure 20 or use Equation (5.12) to extend to higher concentrations. Then use Equation (8.2) with an assumed value of $\tau_d$ ($0{\cdot}08\,\mathrm{N\,m^{-2}}$) to calculate the instantaneous rate of deposition of suspended cohesive sediment from flowing water.

*No field measurements*

6. Use with caution the average settling velocity relationship in Equation (5.12) with an assumed value of $\tau_d$ ($0{\cdot}08\,\mathrm{N\,m^{-2}}$) in Equation (8.2) to calculate the instantaneous rate of deposition of suspended cohesive sediment from flowing water.

7. Estimate the shear stress for deposition for each class of a distributed sediment from Figure 27.

*Example 8.1. Deposition from a current*

To calculate the mass of mud deposited in one hour on a unit area of the bed of an estuary in water at 10°C, 35 ppt

($\rho = 1027\,\text{kg}\,\text{m}^{-3}$, $\nu = 1.36 \times 10^{-6}\,\text{m}^2\,\text{s}^{-1}$) when a current is flowing, obtain estimates of the following parameters:

| | | |
|---|---|---|
| near-bed concentration of suspended mud | $C_b$ | $3\,\text{kg}\,\text{m}^{-3}$ |
| depth-averaged current speed | $\bar{U}$ | $0{\cdot}2\,\text{m}\,\text{s}^{-1}$ |
| water depth | $h$ | $5\,\text{m}$ |

Calculate settling velocity as in Example 5.1:

| | | |
|---|---|---|
| near-bed median settling velocity | $w_{50}$ | $2{\cdot}42 \times 10^{-3}\,\text{m}\,\text{s}^{-1}$ |

Calculate bed shear-stress following the methodology in Example 3.1:

| | | |
|---|---|---|
| bed shear-stress | $\tau_0$ | $0{\cdot}0509\,\text{N}\,\text{m}^{-2}$ |

Calculate or estimate the threshold shear-stress for deposition: In this example take the default value:

| | | |
|---|---|---|
| | $\tau_d$ | $0{\cdot}08\,\text{N}\,\text{m}^{-2}$ |

Use Equation (8.2) to calculate deposition rate, or use SandCalc–Mud–Deposition–Currents–Krone:

| | | |
|---|---|---|
| deposition rate | $dm/dt$ | $0{\cdot}00264\,\text{kg}\,\text{m}^{-2}\,\text{s}^{-1}$ |

Over one hour the mass deposited will be:

| | | |
|---|---|---|
| mass per unit area $= 3600 \times 0{\cdot}00264$ | $m$ | $9{\cdot}50\,\text{kg}\,\text{m}^{-2}$ |

At slack water, the mass deposited could be calculated from Equation (8.1) or by SandCalc–Mud–Deposition–Still Water, and in one hour a mass per unit area of $26{\cdot}2\,\text{kg}\,\text{m}^{-2}$ would be deposited.

## 8.4. DEPOSITION IN WAVES

### Knowledge

Deposition of mud under waves will occur if the peak bed shear-stress, $\tau_w$, exerted by the waves falls below the threshold value for deposition, $\tau_d$. The rate of deposition, $dm/dt$, follows an empirical equation similar to Equation (8.2), but with the current-induced bed shear-stress $\tau_0$ replaced by $\tau_w$:

$$dm/dt \begin{cases} = -(1 - \tau_w/\tau_d)C_b w_{50} & \text{for } \tau_w > \tau_d \\ = 0 & \text{for } \tau_w \leqslant \tau_d \end{cases} \tag{8.4}$$

A sediment with a range of settling velocities can be modelled in an analogous fashion to that described in Section 8.3.

### Procedure

The methods to calculate the rate of mud deposition beneath waves are analogous to those described in Section 8.3 *Procedure*. However, the estimation of $\tau_0$ is replaced by the estimation of $\tau_w$. Methods to calculate $\tau_w$ are described in Section 3.3.

For combined wave and current forces the deposition can probably be estimated using Equation (8.4) with $\tau_{max}$, Section 3.4, in place of $\tau_w$. However, to the Authors' knowledge there are no data available to validate such an approach.

*Example 8.2. Deposition under waves*
Calculate the mass of mud eroded if waves instead of a current are acting under the conditions specified in Example 8.1. In addition to the parameters used in Example 8.1, obtain examples of:

| | | |
|---|---|---|
| significant wave height | $H_s$ | 0·3 m |
| zero-crossing period | $T_z$ | 2·0 s |

Calculate bed shear-stress produced by waves, by the same method as Example 3.2.

| | | |
|---|---|---|
| bed shear-stress amplitude | $\tau_w$ | 0·0450 N m$^{-2}$ |

Use Equation (8.4), or SandCalc–Mud–Deposition–Waves, to calculate:

| | | |
|---|---|---|
| deposition rate | $dm/dt$ | 0·00318 kg m$^{-2}$ s$^{-1}$ |

After one hour the mass deposited will be:

| | | |
|---|---|---|
| mass per unit area = 3600 × 0·00318 | $m$ | 11·4 kg m$^{-2}$ |

# Consolidation

# 9. *Consolidation*

## 9.1. GENERAL

Once the sediment has settled on to the bed because it is in a loose state (partially- or under-consolidated) there is a tendency for self-weight consolidation to take place. This process leads to the sediment density increasing through time and the voids ratio decreasing as trapped pore water is expelled from the settled flocs as they are compressed and from the voids between flocs and other grains. The increase in density is generally associated with an increase in the strength of the bed and the resistance of the bed to erosion. The consolidation process leads to the formation of a density profile within the bed. Under conditions of no further sediment input to the bed surface the consolidation process gives rise to a reduction in the level of the bed. Where the processes of erosion and deposition are continually operating the state of consolidation dictates how the bed will respond to the next period of erosion or deposition. The main elements of the approach have been derived from laboratory experiments. The state-of-the-art in these techniques is discussed by Sills (1997) including some of the shortcomings with laboratory experiments.

## 9.2. VARIATION OF MUD DENSITY WITH DEPTH AND TIME

### *Knowledge*

There is a lack of knowledge regarding the interface between a flowing suspended sediment and a deposited bed. This is due primarily to the difficulty of measuring the flow and suspended-sediment concentration at the necessary resolution (less than

10 mm) to define the physics close to the bed. As a deposited cohesive sediment bed is formed the structure changes from that of a high concentration of suspended sediment in water to a flocculated matrix with a real structure, referred to as the *gel point* (Dade and Nowell, 1991; Mitchener *et al.*, 1996b). From laboratory studies it may be assumed that the sediment deposited to the bed, forming a matrix structure, has an initial formation dry density $C_M$ in the range 20–200 $kg\,m^{-3}$. This value is dependent on the cohesiveness of the sediment and the depositional environment.

The deposited cohesive sediment will consolidate over time under its self weight (Berlamont *et al.*, 1993; Teisson *et al.*, 1993). This process comprises the expulsion of the pore water with accompanying large strains. The excess pore pressures within the cohesive bed will dissipate with time and the interparticle stress $\sigma'_v$ will increase. The interparticle stress $\sigma'_v$, or vertical effective stress, is given by the difference between the total stress and the pore water pressure. The floc matrix will compress to form a structure of higher density with a correspondingly lower permeability $K_I$.

The consolidation process for a cohesive sediment bed will generally result in a density profile which increases with depth below the surface (analogous to Figure 15) and with time. The density at any point and time within a cohesive sediment bed will depend to a large degree on the time history of the deposition to the bed and the physical characteristics of the cohesive sediment.

Two empirical relationships have been commonly derived from laboratory tests in consolidation columns on cohesive sediments (e.g. Ockenden and Delo, 1991); (1) a relationship for the effective stress and (2) a relationship for the permeability. However, even for the same type of cohesive sediment the relationships are dependent on the rate and quantity of sediment deposition. The effective stress $\sigma'_v$ can be expressed as a function of dry density $C_M$ by the following Equation (9.1).

$$\sigma'_v = c_1 + c_2 C_M + c_3 C_M^2 \tag{9.1}$$

A typical relationship is shown by the solid line in Figure 28 with the range of experimental scatter indicated.

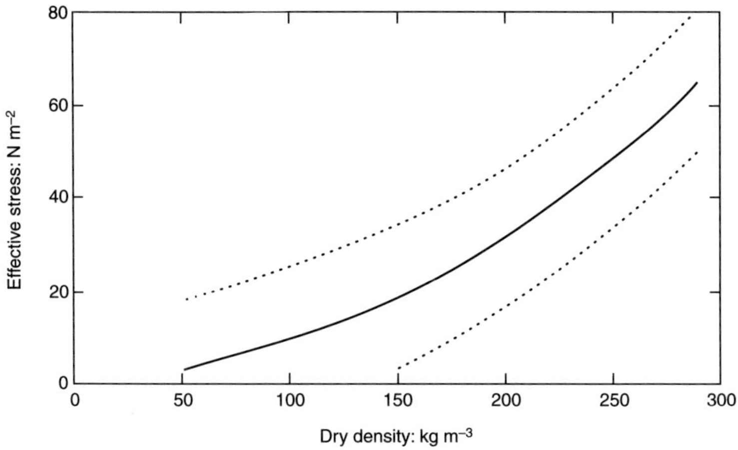


*Figure 28. Typical relationship for the effective stress as a function of bed dry density. Original data obtained from laboratory consolidation tests measured at various times and depths in the bed – the dashed lines indicate the limits of scatter in the experimental data (from Delo and Ockenden, 1992)*

The permeability $K_I$ of the cohesive sediment can also be expressed as a function of dry density $C_M$ (Figure 29) by

$$\log(K_I) = c_4 + c_5 C_M \qquad (9.2)$$

where $c_1 \ldots c_5$ are dimensional coefficients.

Both relationships describe consolidation after the first few hours. In the early stages of consolidation, the coefficients appear to show some dependency on time. The form of these empirical equations may vary (e.g. Table 8) and sometimes the calculations have been made using the voids ratio $V_R$.

Some typical values of permeability at $C_M = 100\,\text{kg}\,\text{m}^{-3}$ are quoted in Table 8.

To predict the density at a given time and depth within a cohesive-sediment bed requires a knowledge of the time history of the deposition to, and erosion from the bed, and the initial formation density of the deposit. The empirical relationships between effective stress and dry density, and permeability and dry density are then used as inputs to a constitutive model for the soil behaviour (e.g. Delo and Ockenden, 1989; Teisson *et al.*, 1993).

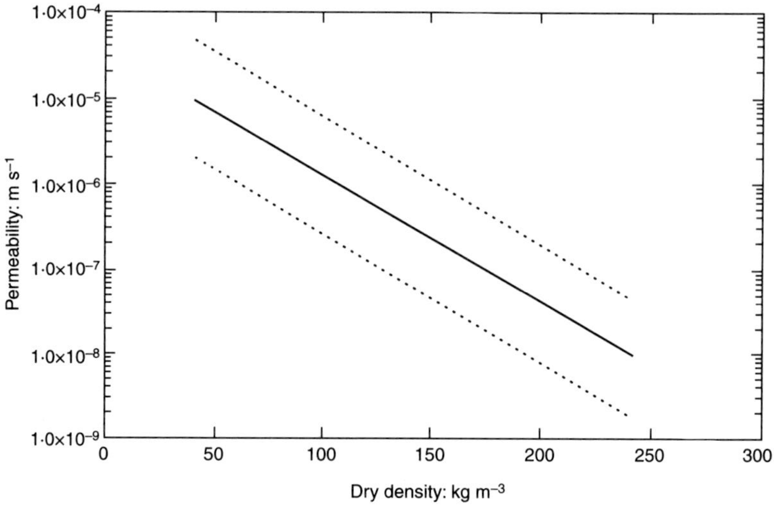


*Figure 29. Typical relationship for the permeability of a mud bed as a function of dry density. Original data obtained from laboratory consolidation tests measured at various times and depths in the bed – the dashed lines indicate the limits of scatter in the experimental data (from Delo and Ockenden, 1992)*

The density structure of cohesive-sediment beds varies considerably between sites and at any particular site. Processes such as periods of exposure to climatic conditions of near-shore intertidal mudflats may also result in desiccation or swelling of the mud matrix (see Section 12), which will further modify the density of these deposits. Laboratory tests have shown however that the suspended solids concentration and salinity – above 2–3 ppt – do not greatly affect the formation density of muddy deposits.

In areas of high deposition subject to regular maintenance dredging, the dry density of underconsolidated deposits could increase from 200 kg $m^{-3}$ near the surface to 500 kg $m^{-3}$ at a depth of 1 m. Typical surface dry densities of intertidal mudflats are in the range 500 to 1000 kg $m^{-3}$. Underneath the recent deposits the dry density – especially in clays – may be as high as 1000 to 1600 kg $m^{-3}$ at a depth of 1 m.

## Procedure

The consolidation behaviour of a mud bed can be determined as follows.

*Field and laboratory measurements*

1. Obtain in-situ measurement of sediment density in the upper 1 m of the cohesive sediment bed at the site of interest using an appropriate in-situ field density probe (e.g. piezo-electric, gamma ray).
2. Conduct consolidation column tests by depositing a cohesive sediment bed from a suspension at an appropriate rate, and measure density and permeability. Determine the values of the constants in Equations (9.1) and (9.2) and the formation density. Note, the laboratory consolidation column-tests can suffer from appreciable pore water drainage between the mud and the inside wall of the column. Care should be exercised in the execution of such tests and in interpreting the results (Sills, 1997).
3. If cohesive sediment from the site has been previously tested in the laboratory use the empirical relationships given, for example in Table 8.
4. Calibrate a consolidation model against measurements and then predict the density with time and depth using the laboratory-derived relationships. Compare model results with the existing conditions from in-situ measurements.

*No field measurements*

5. Use the empirical relationships in Equations (9.1) and (9.2) and previously obtained values of the coefficients with caution to run a mathematical model of the consolidation process to predict the density with time and depth. Assume an initial bed density of:

$$C_M = 70\,\mathrm{kg\,m^{-3}} \tag{9.3}$$

Figures 27 and 28, and Table 8, indicate typical values and ranges of the consolidation parameters for muds tested in the laboratory at HR Wallingford.

*Fluid mud*

For the consolidation of a fluid mud bed the following simple approach can be used. If the shear stress between a moving fluid-mud layer and the bed falls below $0{\cdot}1\,\mathrm{N\,m^{-2}}$ it will dewater (Figures 25 and 26). Fluid mud dewaters at a rate of approximately $0{\cdot}05\,\mathrm{mm\,s^{-1}}$ to form a weak soil with a dry density of

$100\,kg\,m^{-3}$ to $300\,kg\,m^{-3}$ which does not flow easily (HR Wallingford, 1991).

## Procedure

For fluid mud the following approach can be taken:

### *Measurements*

1. Determine the dewatering rate in the field or in a large scale test. Tests in small diameter laboratory columns are not necessarily representative.
2. Use a combination of techniques: e.g. a dual frequency echo sounder to detect the fluid mud-water interface and fluid mud-bed interface, water sample collection, current metering above and in the fluid mud layer, and a density probe to measure the bed density profile.

### *No new data*

3. Use the relationships depicted in Figures 28 and 29 to estimate the consolidation of a fluid mud.

# Mixtures of mud and sand

# 10. Mixtures of mud and sand

## 10.1. GENERAL

Many natural sediments comprise a mixture of different particle sizes and because of the interaction between these different fractions the mixture behaves in a different way than the constituent parts. This is especially true of mixtures of (cohesionless) sand when it is mixed with an amount of cohesive mud or clay particles.

## 10.2. EROSION

### *Knowledge*

There are three main components to mixed-sediment erosion behaviour:

(*a*) $\tau_e$, the critical applied shear-stress for the initiation of erosion (units $N\,m^{-2}$)
(*b*) the mode of erosion, whether the mixture behaves as a mud or a sand and is transported as bedload or suspended load
(*c*) $dm/dt$, erosion rate ($kg\,m^{-2}\,s^{-1}$) once $\tau_e$ has been exceeded.

The erosion of mixed sediments has been investigated both with homogeneous mud beds tested in the laboratory or on cores retrieved from the field for testing in the laboratory (Figure 14), or as natural sediments eroded by field instruments (e.g. Black and Paterson, 1997).

Laboratory flume tests (in straight or annular flumes) involve mixtures which have been mechanically stirred-up and redeposited in the laboratory environment, and these generally represent underconsolidated beds with a less well-developed sediment matrix structure than those found in natural estuaries and mudflats.

Natural sediments are often layered, and contain varying amounts of biological organisms and secretions, which further affect the erosion behaviour (as discussed in Section 2.4).

### 10.2.1. Threshold erosion shear-stress ($\tau_e$)

#### *Increase in $\tau_e$ with sand added to mud*

Sand increases the binding between the clay particles and results in a more compact and dense matrix, which is then more resistant to erosion (e.g. Figure 30). For underconsolidated muds adding sand to mud increases its erosion resistance (thus increasing $\tau_e$) because of the increased bed-density and influence on consolidation rates (Section 10.3). Laboratory tests with homogeneously mixed beds (Mitchener *et al.*, 1996a) showed that adding up to 50% sand by weight to a mud could increase $\tau_e$ by up to a factor of 2. This result was borne out by in-situ field measurements at the Sellafield mud patch in the Irish Sea (Feates and Mitchener, 1996).

Adding sand to mud increases the bed density via compaction because of the following:

- increased self-weight consolidation
- increased drainage
- higher solids fraction within the sediment due to the larger sand grains.

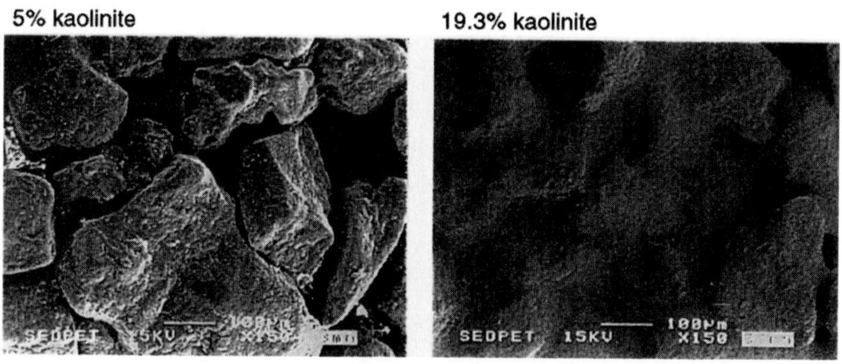


*Figure 30. Scanning electron microscope plates of kaolinite/sand mixtures, with 5% and* 19·3% *fines respectively, showing the binding effect of the clay network structure. Reproduced from Mitchener* et al. *(1996a), Erosion of mud/sand mixtures,* Coastal Engng, ***29****, 1–25, with permission from Elsevier Science*

#### *Increase in $\tau_e$ with mud added to sand*

Small amounts of clay added to a sand can increase significantly its erosion resistance. The addition of mud to a sand creates a cagework structure around the sand grains trapping the sand within a clay aggregate envelope. Figure 30 shows electron microscope plates of kaolinite and sand mixtures, showing the binding cagework of kaolinite at 19·3% mud content on the right, where the mud matrix fully encompasses the sand grains. The erosion resistance is then the combined result of the sand gravimetric properties and the binding influence and density of the mud cage. Alvarez-Hernandez (1990), Torfs (1994), Kamphuis and Hall (1983), Williamson and Ockenden (1993) and Panagiotopoulus *et al.* (1997) have all conducted laboratory flume tests which led to the general conclusion that additions of cohesive material to a sand bed increase the erosion resistance.

The sand only (0% mud) threshold, $\tau_{cr}$, is governed by the sand properties (Soulsby, 1997). The increased erosion resistance with added clay minerals is governed by the added mud characteristics. For example, kaolinite has been found to be more binding than montmorillonite (Torfs, 1994).

A general maximum appears to occur where about 20 to 30% mud is added to sand; the value of $\tau_e$ can increase by as much as a factor of 10 times over the value of $\tau_{cr}$ for pure sand (Mitchener *et al.*, 1996a). This is depicted schematically in Figure 31. The maximum value will depend on the grain size and porosity of the sand and the density and cohesiveness of the interstitial clay and values of 3 to 5 may be more typical.

$\tau_e$ can also increase with the following:

- plasticity index (Kamphuis and Hall, 1983)
- organic content (Young and Southard, 1978)
- biological activity (Paterson *et al.*, 1990).

### *10.2.2. Mode of erosion*

There are three modes of erosion for homogeneously mixed beds:

(*a*) non-cohesive erosion, predominantly as bedload
(*b*) transition zone of mixed mud/sand type erosion, sometimes with small ripples
(*c*) cohesive type erosion, predominantly as suspended load.

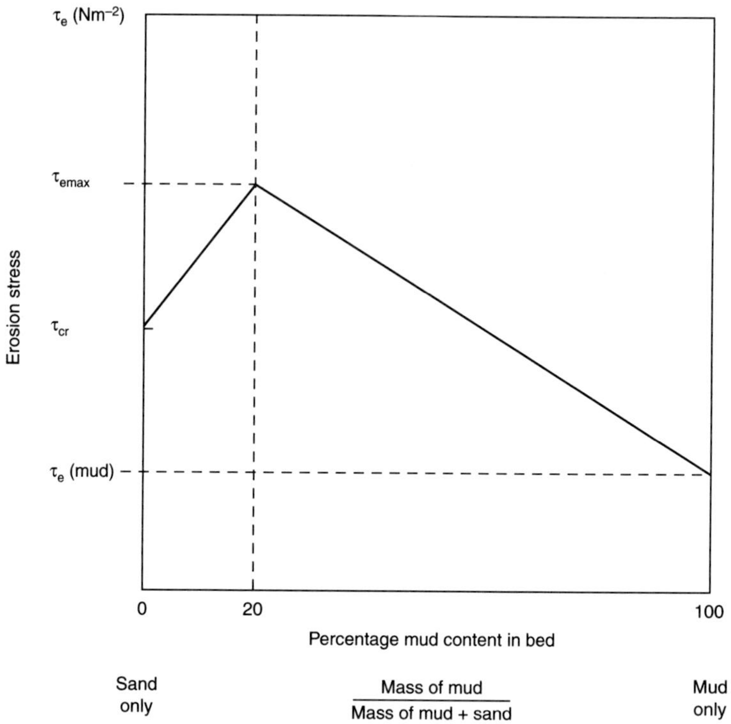


*Figure 31. Representation (solid line) of the effect of the proportion of mud to sand deposits on the threshold shear-stress for erosion $\tau_e$. The values of $\tau_e$ (mud) and $\tau_{cr}$ denote the values for mud and sand only (after Chesher and Ockenden, 1994)*

Various researchers (Dyer, 1986; Raudkivi, 1998) have suggested that a clay content of 5–10% by dry weight is sufficient to control the sediment properties such that they behave like a clay with an increased resistance to erosion. Panagiotopoulos *et al.* (1997) suggest that when the clay mineral fraction is greater than 11–14% by weight the sand grains are no longer in contact with each other and the mixture is controlled by the clay structure. The transition zone is dependent on the sand and clay characteristics:

- for kaolinite the transition is at about 3% of added clay by weight (Mitchener *et al.*, 1996a)
- for montmorillonite the transition is 7–13% of added clay by weight (Mitchener *et al.*, 1996a)

- for Laponite clay the transition is 5–15% of added clay by weight (Alvarez-Hernandez, 1990).

In general these results confirm the oft quoted rule of thumb that cohesive sediment transport equations for resistance to erosion should be used where a sediment mixture contains in excess of 10% fine particles, < 63 μm in size (Mitchener *et al.*, 1996a) (Section 2.3).

The size of aggregates eroded from a muddy bed decreases with increasing amounts of added sand.

### 10.2.3. Layered beds

For layered beds erosion should be modelled as sequence of sandy and muddy type erosion events with appropriate threshold values and erosion rates and modes of transport for the different layers.

The erosion of muddy layers in a mixed sediment is governed to some extent by the erosion of the sandy layers: if the flow regime is sufficient to erode the sand, then the movement of sand can initiate and enhance the muddy layer erosion. The presence of sand particles being transported in the overlying water column can significantly increase the erosion. Kamphuis (1990) found that even 0·1% sand by volume in the water can increase the erosion depth (thus erosion rate) by a factor of 5.

If granular material is deposited over a cohesive layer, then if the bottom layer of grains in the granular layer is stable, the mud will not erode (Kamphuis, 1990).

### 10.2.4. Erosion rate $(dm/dt)$

Once the percentage of fine sediment is above the transition for the mode of erosion (Section 10.2.2) the erosion rate can be reduced by as much as a factor of 5 compared to the case of sand only (kaolinite tests, Torfs, 1994).

It has been shown (Appendix 1 of HR Wallingford, 1990) that the presence of sand in the fluid-sediment matrix has no significant effect on the rheological parameters of a non-Newtonian mixture for concentrations less than 20% by volume. This has implications for the fluidisation behaviour of mud beds reviewed earlier.

*Procedure*

1. Measure $\tau_e$ in situ if possible at a number of different sites, using an in-situ erosion instrument. See Black and Paterson (1997) for a review of in-situ measurement devices.
2. Establish the structure of the mixture, whether it is homogeneous or layered, and treat it as a stack of horizontal layers of appropriate thickness if it is layered. Apply appropriate erosion equations to model the behaviour of each layer.
3. If no erosion measurements have been made take sediment samples and establish their density and sand content. Establish the mud and clay content, and then apply the following approximations:

   (*a*) calculate $\tau_{cr}$ for sand only (Soulsby, 1997)
   (*b*) calculate $\tau_{emax}$ at 20% mud as in Chesher and Ockenden (1994) – see Figure 31:

   $$\tau_{emax} = 5 \times \tau_{cr}$$

   The value of the multiplier for $\tau_{cr}$ will vary with the sediment properties.
   (*c*) define $\tau_e$ for mud only (from measurement, or use an approximation from Table 8, or if there is no available data use 0·1–0·2 $Nm^{-2}$ for recently deposited sediment).

   Then use the approximation (Figure 31) that $\tau_e$ increases linearly between $\tau_{cr}$ for sand and $\tau_{emax}$ in the region 0% to 20% mud content, and then decreases linearly between $\tau_{emax}$ and $\tau_e$ in the region 20% to 100% mud content.
4. For the mode of erosion apply the following:

   (*a*) below 5% mud, allow the sediment to erode in a non-cohesive manner
   (*b*) for 5% to 15% mud, the sediment is in a transition zone for the mode of sediment transport behaviour
   (*c*) above 15% mud, allow the sediment to erode as if it were a mud. Reduce the erosion rate by a factor of 5 (Torfs, 1994).

## 10.3. DEPOSITION AND CONSOLIDATION

### *Knowledge*

Mixed sediment beds occur in two different forms:

- homogeneous
- layered.

Sand and mud deposit independently from the water column. Sand deposits very quickly, whereas fine muddy material may take hours if not days to settle. The deposition and resulting composition of the bed will depend upon the degree of flocculation and hence settling velocity of the muddy material.

A single input of sediment into the water column will result in a segregated bed, with a coarser, denser lower layer and a finer, less compact upper layer, which is sand-free.

For single input mixed-beds the following features have been observed:

- segregation occurs even if no sand is present
- increasing the sand content increases the segregation
- increasing the sand content increases the overall bed density
- increasing the sand content results in higher consolidation rates (owing to compaction and the passage of sand grains through the mud matrix)
- higher sand contents result in beds with higher permeabilities (Williamson and Ockenden, 1993).

For a continuous sediment input from the water column, such as in a typical tidal influx, the structure of the resultant mixed bed is dependent on the input rate and concentrations, and the sand content. These factors affect the development of a cohesive matrix structure, or *gel point* which then governs the trapping of falling sand grains and the extent of segregation.

(*a*) The *gel point* is established if the input rates are low. The descending clay particles have sufficient time to become interconnected, and form a network structure, which traps sand grains homogeneously within the consolidating matrix.

(*b*) Hindered settling and segregation occurs when the input rate is high. The sand grains then fall freely through a high mud concentration layer and form a distinct segregated layer at the base.

The segregation of sediment depends on the following factors:

- type and cohesiveness of the mud
- mineralogical and chemical composition
- input sediment density
- sand content
- input rate.

Once the bed structure is established, sand grains are held within the sediment matrix.

The *gel point* occurs at a dry mass sediment concentration of 80–180 kg m$^{-3}$ (corresponding to an excess density $\rho_B - \rho$ of about 50–110 kg m$^{-3}$) depending on the sediment type. Increased amounts of non-cohesive material in the settling suspension appears to increase the *gel point* concentration by limiting the space for the mud network to fill (Torfs *et al.*, 1996).

In summary, increasing the sand content within a bed will:

- increase the overall density
- reduce the thickness of the bed by increasing the consolidation rate (with a maximum rate at about 30% sand by weight)
- increase the final bed density (up to a maximum at about 20–30% sand by weight)
- increase the permeability.

Layered beds result when sequential deposition and erosion events occur. The resultant beds have stepped density profiles with sharp density peaks at the base of each layer corresponding to sandy layers. More information on the deposition of mud/sand mixtures can be found in Torfs *et al.* (1996).

### *Procedure*

The procedure for predicting the deposition and consolidation of mud-sand beds is as follows:

1. For a depositing bed, split the sediment into grain-size bins (e.g. Equation (8.3)), each with an appropriate settling velocity.

2. Allow the deposited bed to be comprised of layers of different grain-size material wherever the concentration is less than the *gel point* (typically 150 $kg\,m^{-3}$).
3. Assume the depositing bed is mixed where the mud concentration is over 150 $kg\,m^{-3}$.
4. Predict the increase in bed density with time and depth using Equations (9.1) and (9.2) or allow a (linear) increase in density vertically down through a bed, with the rate of increase determined from measurements of density at different levels in the bed.
5. Allow the presence of sand within a mixed matrix to increase the consolidation rate, with a maximum at 30% by weight of sand in the bed.

# Mathematical modelling

# 11. Mathematical modelling

## 11.1. GENERAL

While the flows that occur in estuaries and coastal situations are frequently very repetitive, for example with similar current speeds on each spring tide, the rate of sediment transport may be a great deal more variable. It depends on the currents and on the waves present, features such as wind and human activity (including navigation and dredging) and most importantly on the sediment properties. For these reasons, sediment transport modelling prediction is as yet at a much lower level of certainty than the modelling of waves and flows.

Sediment transport models can be classified according to their order of dimensionality, for example:

- 3D
- 2DH (two dimensions in the horizontal)
- 2DV (two dimensions in the vertical)
- 2D 2 layer
- 1D
- point models
- particle (Lagrangian) models.

For the prediction of sediment transport it is not only important to understand the hydrodynamics which are driving the sediment transport, but it is also important to adequately represent the physics of the sediment transport mechanism. Inaccuracies in the representation of the hydrodynamics will lead to uncertainties in the magnitude of the sediment transport and in the identification of the most important sediment transport mechanisms. However, inadequate representation of the physical properties of the sediment itself may lead to greater uncertainties.

Ideally, before undertaking computational modelling of sediment transport, measurements of the sediment distribution and properties at the site should be obtained. Procedures for obtaining information on various parameters have been discussed throughout this manual. In the absence of detailed site-specific information it is possible to make useful predictions of sediment transport by applying some sensible assumptions on the nature and properties of the sediment and to use this as a basis for a series of sensitivity tests. Indeed, this is an approach which can help to identify the requirements for a field monitoring exercise. However, the application of appropriate assumptions needs to be undertaken with care. It is dangerous to rely too heavily on the results of sediment transport models, without backing this up with a sound physical understanding of the processes at work.

### *Procedure*

Computational modelling of mud transport is best used as a part of an overall assessment procedure which also includes desk based study and appropriate field investigations. An initial desk based study should assess the main mechanisms for mobilising sediment and for transport of sediments and should identify the possible supply of sediment, either in the form of existing deposits which can be mobilised or as suspended sediments carried into the area of interest from elsewhere.

Once the main mechanisms are known, a suitable modelling approach can be selected which can represent these mechanisms. This is a summarised version of earlier guidance given by Cooper and Dearnaley (1996).

To help formulate a suitable approach the following questions should be answered.

### *Forcing for mud model*

- Is the mud moved primarily in suspension? If so, information on currents are required, usually in the form of results of a computational flow model, compatible with the mud model to be used.
- How many dimensions are required? Is the water column stratified? Is there vertical variation of residual currents? Are there strong inputs of fresh water (estuaries)?

- Are the fluid mud layers produced by hindered settling or wave fluidisation? Does the model need to represent these fluid mud forming processes?
- Are waves important in resuspending or fluidising the mud?
- How many tide types/fresh water flow conditions/wave conditions should be used?

*Initial conditions*
- What is the initial distribution of sediments on the bed? Where are there deposits of mud which can be eroded?
- What are the suspended sediment concentrations? Ideally access to measurements is required for determining the initial conditions.

*Boundary conditions*
- What are the suspended sediment concentrations at the boundaries of the model?

*Parameters describing sediment properties*
Typically, models of mud transport involve the following parameters which the user must provide:

- settling velocity
- erosion rate
- critical shear-stress for erosion
- critical shear-stress for deposition.

The advice given in this manual can be used to pick values for these.

Finally, it is extremely valuable to have data for calibration or verification of the model predictions.

## 11.2. APPROACHES TO USE OF A MODEL

If adequate field data are available to define sources of mud (inputs from rivers, model boundaries, areas of erosion) as well as information on suspended sediment concentrations and siltation rates, it may be possible to calibrate the model by adjusting the sediment parameters and the locations and amounts of sediment sources. The model can then be used in a predictive mode to look at the effects of engineering works, for example. Note that the model will only give sensible results if the important sediment transport mechanisms are represented. For example, if a harbour suffers siltation during storms because waves are greatly increasing suspended sediment concentrations, a model which represents only transport by currents in calm conditions will not give sensible results.

In the absence of a comprehensive field data set, models can be used to assist in examination of different processes and to test out hypotheses by setting up schematic tests. For example, the rate of siltation in a harbour basin could be investigated by imposing a fixed suspended-sediment concentration at the harbour entrance and using the model, linked with a suitable hydrodynamic model, to calculate the rate of input and settling of sediments. More extensive guidance on making harbour siltation calculations is provided by Dearnaley *et al.* (1997).

Advice on how to deal with a wide range of currents and wave conditions is given in Soulsby (1997, Chapter 12). Use of representative tides and wave conditions is possibly more difficult in the case of muddy sediments than for sand, because of the additional timescales introduced by the consolidation of sediments and low settling velocities, which means that the effect of events such as storms can be more persistent than for sand. However the same principles can be applied, although perhaps with a greater factor of uncertainty.

# Intertidal processes

# 12. *Intertidal processes*

## 12.1. GENERAL

Many of the basic premises derived for the sediment dynamic, biological and chemical behaviour of sub-tidal cohesive sediments are applicable to intertidal sediments. However, the muddy intertidal zone (mudflat) of estuaries and coastal areas is subjected to a range of processes which modify the basic behaviour of the sediment and the way in which it is eroded, transported and deposited. The chemistry, biology and sediment dynamics of mudflats has many complex interactions, and some real advances have been made in recent years especially with the advent of a number of multi-disciplinary research projects (for example, the EC/national funded PRO-MAT and INTRMUD, the Canadian LISP and the British LISP-UK). The results of these projects are now finding their way into the literature (e.g. Paterson, 1994; Black *et al.*, 1998; Dyer, 2000).

## 12.2. FACTORS FOR CONSIDERATION

The main factors characterising mudflats are as follows. They are subjected to:

- tidal wetting and drying under the action of the flood and ebb tidal fronts
- (very) shallow water processes and the action of waves advecting back and forth across the mudflat with the tide
- exchanges of water and sediment with the adjacent saltmarsh and the nearshore region
- subaerial exposure and meteorological controls on the sediment properties
- the activity of flora and fauna which can act either to stabilise or destabilise the sediment (Section 2.4.1).

*Table 10. Summary of processes influencing intertidal sediment and the effect on the sediment properties*

| Process | Effect |
|---|---|
| Tidal wetting/ drying | The tidal wetting and drying leads to a translation of water across the intertidal profile plus associated changes in water pressure and exchanges with the groundwater in the sediment. The speed of the tidal bore can be quite fast and associated with high sediment concentrations.<br><br>The movement of the tide also produces a sweeping motion of the surf zone across the mudflat and the influence of wave activity can reach different parts of the mudflat. |
| Evaporation/ drainage | Evaporation leads to a decrease in the water content of the surficial sediment and an increase in the salinity of the pore water. Drainage through the sediment leads to a decrease in the water content of the sediment. A reduced water content results in a sediment with increased strength and hence resistance to erosion (providing it is mechanically intact). |
| Dessication | The intense drying out of the surficial sediment results in a baked surface layer with very little residual water content, and hence strength. Dessication polygons are likely to result where the sediment has shrunk and the sediment is possibly less resistant to erosion by the incoming tide. |
| Rainfall | Rainfall can have a number of influences on the surface sediment. If the surface is wet and the moisture storage is saturated the rainfall is likely to contribute to runoff downslope or preferentially along channels or gullies. If the sediment is dry the rainfall will recharge the moisture levels in the sediment, unless the rainfall exceeds the rate at which it can infiltrate into the sediment in which case runoff will occur. Soil particle detachment can be by rainfall impact or by runoff.<br><br>Rainwater will also reduce the salinity of the surface sediment and this can lead to deflocculation and a reduction in shear strength. Depending upon the prevailing conditions the shear strength can be built up again during the remaining exposure period. |
| Ice | The formation of ice crystals in the sediment leads to a mechanical reduction in the strength of the sediment and an increased susceptibility to erosion by waves and currents.<br><br>Blocks of ice formed on the mudflats can be floated off by the rising tide allowing sediment to be transported to another location depending on the current circulation pattern. |

The processes acting on intertidal sediments that differentiate them from sub-tidal deposits are listed in Table 10. The wetting and drying processes are likely to have a more pronounced influence on the sediment if it contains one of the swelling clay minerals that contain adsorped water (e.g. montmorillonite). The moisture content and pore water salinity of the sediment plus the prevailing temperature and light level will determine the level of biological activity in the sediment and on the surface.

The mudflat environment is characterised by the translation of the tide sweeping across the intertidal profile with associated high initial velocities in the shallow tidal front. The influence of wind waves on the intertidal sediments will also be felt across the mudflat as the tide produces a translation of the surf zone across the profile. The relative magnitudes of the cross-shore and long-shore components of the current will be controlled largely by the tidal range, and the characteristics of the tidal wave in the nearshore zone and the steepness of the mudflat (Whitehouse and Roberts, 1999). On the larger scale of the estuary-mudflat-saltmarsh system, exchanges of water and sediment take place between these three zones which are governed by tidal and wave processes. In the estuary situation the supply of sediment to the mudflat can vary on a fortnightly (spring-neap) cycle as sediment is released from fluid mud deposits in the estuary by the increased stirring of the spring tides, becoming available for deposition on the mudflat. Whether the sediment remains on the mudflat and consolidates into a resistant layer depends upon the sequence of events after the deposition has taken place (usually at slack water). The influence of waves can be dramatic with waves only 6 cm high causing erosion of the sediments at the edge of the surf zone.

Even if the waves do not erode deposited sediment, the increased wave stirring in the surf zone ensures a high concentration of sediment is maintained in this narrow strip. Hence on the ebbing tide fresh deposits can be resuspended by the current or by the waves and advected off the mudflat with the receding tide. Therefore, the timing of wave action relative to the height of the tide is likely to control whether any sediment remains deposited on the mudflats. If erosional conditions occur when there is only a thin layer of protective mud on the surface of the flat this can allow erosion of the underlying stratum. This may contribute to a progressive recession of that shoreline despite the periodic deposition of sediment on the mudflat profile. The rate at which the net build up of sediment takes place is a relatively slow (seasonal) process and it is only possible to determine whether a mudflat is eroding or accreting by looking at long time-series of shoreline position data (e.g. greater than 20 years duration). By correlating bed level data with available information on the prevailing environmental conditions and processes it should be possible to assess whether this trend is likely to continue at least in the short-term.

Mudflats can act either as a source or sink for fine-grained sediments depending upon variations in tidal range, sediment supply, wave climate and biological activity, plus the frequency and timing of storm waves and other climatic factors. The mudflat can be characterised by its cross-shore profile (convex or concave, or a composite of these), by variations in the thickness and properties of the mobile sediment in a shore-normal direction, and the presence of morphological features of various shapes and sizes (Whitehouse *et al.*, 2000). Mudflats are often very three-dimensional in the longshore direction and the thickness of the lens of recently deposited sediment and bedform type can vary on an annual basis (O'Brien *et al.*, 2000). There is some debate as to whether the cross-shore profile of the mudflat is deterministic of the prevailing environmental conditions and in an attempt to address this issue a typology for mudflats has been suggested by Dyer (1998a). Further research within the INTRMUD project has resulted in a quantitative classification scheme for mudflats (Dyer *et al.*, 2000). Techniques for predicting the cross-shore profile have been developed by, among others, Whitehouse and Roberts (1999) and Roberts *et al.* (2000).

Experience from sites around the UK suggests that wave damping from fluid mud layers on mudflats is not significant as the mud layers do not usually exist in thick enough layers on the mudflat profile. The same is not true of some other microtidal mudflat areas where wave attenuation by fluid mud can be significant (e.g. Lee and Mehta, 1997).

# Case studies

# 13. Case studies

## 13.1. GENERAL

Some case studies are presented here to demonstrate how the *Knowledge* and *Procedure* sections presented in the preceding chapters can be implemented to address typical problems of marine engineering importance. Extensive tables are included to demonstrate how the calculations are performed and how the results can be presented.

## 13.2. INFILL OF HARBOUR BASIN

A new harbour basin is to be constructed in a muddy coastal area. The harbour will be 400 m by 400 m in plan, with a single entrance 50 m wide. The depth of the harbour will be 10 m at mean water level, with a spring tidal range of 4 m and a neap tidal range of 3 m. It is necessary to estimate the amount of maintenance dredging which will be required.

Measurements of suspended sediment concentrations have been made, 1 m above the bed at hourly intervals on a spring tide, at a point close to where the mouth of the new harbour will be. Tidal currents in the main channel reach a maximum of approximately $0{\cdot}5\ \mathrm{m\,s^{-1}}$ on a spring tide.

The amount of sedimentation in the harbour will depend on the amount of water entering the harbour, how much sediment it is carrying and how much of that sediment can deposit before being washed out of the harbour again. Because of the deep water in the harbour, and the shelter provided from waves, the effect of waves can be neglected.

Current speeds in the harbour must be estimated. The tidal volume of the harbour is given by the surface area multiplied by the tidal range, given for spring tides by

$400 \times 400 \times 4 = 640\,000\,\mathrm{m}^3$. The cross-sectional area of the entrance is $50 \times 10 = 500\,\mathrm{m}^2$ at mean water level. With a sinusoidal tide, the rate of change of water level is greatest at mean water level, equal to $\pi R/T$, where $R$ is the tidal range and $T$ is the tidal period, giving $2{\cdot}8 \times 10^{-4}\,\mathrm{m\,s^{-1}}$ in this case. Hence the peak tidal flow through the mouth is $400 \times 400 \times 2{\cdot}8 \times 10^{-4} = 44{\cdot}7\,\mathrm{m^3\,s^{-1}}$. With a cross-sectional area of $500\,\mathrm{m}^2$ at the mouth, this gives a peak cross-section averaged flow velocity of $0{\cdot}09\,\mathrm{m\,s^{-1}}$. The tidal flow speeds within the basin will be smaller than those at the mouth. This gives a bed shear-stress significantly below the thresholds for deposition or erosion, hence it is safe to assume that there will be no re-erosion of deposited sediments.

In addition to the tidal exchange of water between the basin and the external waters, there may also be a contribution from horizontal or vertical circulations. Horizontal eddies are common in the mouth of harbour basins. Their strength and shape are very sensitive to the geometry of the harbour and for a proposed harbour development, where it is not possible to measure the flow, computational or physical modelling is the only way to quantify this type of flow accurately.

Another possibility is a vertical circulation driven by density differences. This can occur when the harbour is in an estuary where there is a horizontal salinity gradient. Because the exchange of water in the harbour is relatively slow, the salinity in the estuary channel can vary much more quickly than the salinity in the harbour, leading to a salinity difference between the harbour and the estuary and hence vertical exchanges of water. Again this is very difficult to quantify without a complex modelling exercise.

As the velocities within the harbour will be so low, it is reasonable to assume that the bed shear-stress within the harbour is zero. The calculated value for a depth of 10 m and the estimated maximum current speed of $0{\cdot}09\,\mathrm{m\,s^{-1}}$ is $0{\cdot}01\,\mathrm{N\,m^{-2}}$. The amount of deposition then depends on the amount of sediment carried into the harbour and the settling velocity.

With an average depth of 10 m, a particle settling at $1\,\mathrm{mm\,s^{-1}}$ will settle from the surface to the bed in 10 000 seconds, equal to just under three hours. There will be a distribution of settling velocities of particles: some will settle faster than others and there will always be a small proportion of unflocculated sediment which settles very slowly. The majority of the sediment will be in the lower half of the water column and so the average distance through which a particle must settle before reaching the bed is generally less than the whole water depth. Therefore a reasonable assumption is that all of the sediment entering the harbour will deposit within the harbour. In practice, as noted above, there will always be a proportion of the sediment which has a very low settling velocity, because some of the particles are unflocculated or contained in very small flocs, so some of the material will remain in suspension and thus not deposit in the harbour. In a case like the one described here, it is likely that only a small proportion of the total sediment mass will fall into this slow settling category.

In a sheltered harbour with slow current speeds, there will be little or no natural resuspension of bed material, but there is the possibility of the propellers and hulls of moving vessels disturbing sediment from the bed. Once in suspension, some of this material may escape from the harbour, but the volumes are likely to be relatively small. More often, vessel movements act to redistribute sediments to the less actively-used parts of the harbour where disturbance is less frequent.

We will assume that the volume exchanged on each tide is equal to the tidal volume, while bearing in mind that in some harbours, this may be significantly enhanced by vertical or horizontal circulations. One aspect of harbour design is to optimise the configuration of the entrance to minimise that type of exchange flow.

In order to estimate the mass of sediment entering a prospective harbour, the concentration in the surrounding waters must be assessed. This can be done by taking water samples prior to harbour construction and carrying out an analysis to measure the sediment concentration. It is assumed that the sediment regime is not significantly affected by the new structure.

We will assume that the measurements shown in Table 11 have been made.

The concentration on the flood phase of the tide will influence the mass of sediment entering the harbour. Assume that the flood covers the six hour period represented by HW + 7 to HW + 12 in Table 11. The rate of water entering the harbour will be proportional to the rate of rise of water level. By assuming a sinusoidal rate of change of water level and by splitting the tidal volume into six hourly parts, the percentage of the tidal volume which enters

*Table 11. Measured suspended-sediment concentrations for spring and neap tides*

| Time | Spring tide: $mg\,l^{-1}$ | Neap tide: $mg\,l^{-1}$ |
|---|---|---|
| HW | 19 | 10 |
| HW + 1 hour | 30 | 16 |
| HW + 2 hours | 32 | 29 |
| HW + 3 | 70 | 44 |
| HW + 4 | 90 | 40 |
| HW + 5 | 86 | 31 |
| HW + 6 | 38 | 15 |
| HW + 7 | 39 | 22 |
| HW + 8 | 73 | 34 |
| HW + 9 | 112 | 32 |
| HW + 10 | 90 | 48 |
| HW + 11 | 43 | 21 |
| HW + 12 | 40 | 12 |

in each hour can be shown to be: 7%, 18%, 25%, 25%, 18% and 7%.

The total tidal volume on a spring tide has been shown to be $640\,000\,m^3$. On a neap tide it is $400 \times 400 \times 3 = 480\,000\,m^3$. Therefore the volumes entering in each hour during the spring tide, and hence the mass of sediment entering in each hour, can be calculated and the results are given in Table 12. Note that $1\,kg\,m^{-3} = 1000\,mg\,l^{-1}$.

The process can be repeated for the neap tide as shown in Table 13.

By approximating a year of conditions as 350 spring tides plus 350 neap tides gives a total annual input of sediment as $22{\cdot}65 \times 10^6\,kg = 22\,650$ tonnes of dry sediment. To work out the depth of deposit, we must assume a value for the density of bed deposits. Taking $1300\,kg\,m^{-3}$ as a bulk density and using Table 2 gives a dry density $C_M = 446\,kg\,m^{-3}$. The area of the basin is $400 \times 400 = 160\,000\,m^2$, so if the deposition is evenly distributed over the whole harbour the dry mass deposited per

*Table 12. Calculated mass of sediment entering the harbour on a spring tide*

| Time | Percentage | Volume: $m^3$ | Concentration: $kg\,m^{-3}$ | Mass: kg |
|---|---|---|---|---|
| HW + 7 | 7 | 44 800 | 0·039 | 1747·2 |
| HW + 8 | 18 | 115 200 | 0·073 | 8409·2 |
| HW + 9 | 25 | 160 000 | 0·112 | 17 920·0 |
| HW + 10 | 25 | 160 000 | 0·090 | 14 400·0 |
| HW + 11 | 18 | 115 200 | 0·043 | 4953·6 |
| HW + 12 | 7 | 44 800 | 0·040 | 1792·0 |
| | | | Total | 49 222·0 |

unit area is 141·6 $kg\,m^{-2}$, which at a dry density of 446 $kg\,m^{-3}$, corresponds to a depth of mud of 0·32 m. In practice, the distribution will tend to be uneven, often with larger values near the entrance, if sediment is depositing relatively quickly after it enters the harbour. Therefore there will be some parts of the harbour where the loss of water depth is more rapid than 0·32 m per year. The currents produced by vessel movements will also serve to redistribute the material.

Another effect which is potentially important is the effect of storms in increasing suspended-sediment concentrations. Large waves, even if they are remote from the site of interest can substantially increase the concentration of fine sediments in the

*Table 13. Calculated mass of sediment entering the harbour on a neap tide*

| Time | Percentage | Volume: $m^3$ | Concentration: $kg\,m^{-3}$ | Mass: kg |
|---|---|---|---|---|
| HW + 7 | 7 | 33 600 | 0·022 | 739·2 |
| HW + 8 | 18 | 86 400 | 0·034 | 2937·6 |
| HW + 9 | 25 | 120 000 | 0·032 | 3840·0 |
| HW + 10 | 25 | 120 000 | 0·048 | 5760·0 |
| HW + 11 | 18 | 86 400 | 0·021 | 1814·4 |
| HW + 12 | 7 | 33 600 | 0·012 | 403·2 |
| | | | Total | 15 494·4 |

water column. Measurements of suspended-sediment concentrations are normally made on relatively calm days, because of the practicality of working from a small boat and so do not generally reflect the full range of concentrations which can occur. If this is likely to be a very important issue, then the possibility of self-contained recording turbidity meters mounted on a bed frame or on a suitable structure should be considered as an alternative source of information on suspended-sediment concentrations.

## 13.3. ACCRETION IN A CHANNEL

This example addresses a similar situation to the example in Section 13.4 in Soulsby (1997), but in this case involving an area of muddy sea bed, rather than a sandy one. An approach channel to a harbour is to be dredged through a muddy bank and the annual maintenance dredging requirement has to be calculated. The channel is aligned at right angles to the coastline, such that tidal currents pass over it at right angles. We will assume initially that wave effects are small and can be neglected. The mean water depth at the shoal is 5 m and the channel is dredged to a mean navigation depth of 10 m, with side slopes of 1:5 and a bottom width of 100 m. Therefore the overall width of the channel is $100 + 2 \times (5 \times 5) = 150$ m.

The spring tidal range is 2 m, tidal currents are equal on ebb and flood, with peak depth-averaged speeds of $0{\cdot}6\,\text{m}\,\text{s}^{-1}$ on neaps and $1{\cdot}0\,\text{m}\,\text{s}^{-1}$ on springs. The sea bed is composed mainly of mud and the dominant mode of sediment transport is in suspension. Water samples have been collected from mid-depth at a point near the proposed channel, at hourly intervals over one spring tidal cycle and one neap tidal cycle.

Possible mechanisms of sediment infill in the channel are the following:

- Direct deposition into the channel during slack water periods from material naturally in suspension in the water column.
- Deposition onto side slopes, and subsequent accumulation in the channel due to sediment sliding down the side slopes.
- Enhanced direct deposition into the channel during periods of wave activity which increase natural background suspended-sediment concentrations.

- Fluidisation of the muddy bed adjacent to the channel during storms and subsequent interception of near-bed mobile layers by the channel.
- Collapse of unstable side slopes into the dredged channel.

We will concentrate on evaluating the first of these.

Based on the conservation of water volume as the current passes over the channel, the depth-averaged current speed over the channel will be approximately half that on the shoal, because the water depth is approximately twice as large, depending on the tidal level. Our approach will be to divide the tide into half hour time-steps, and to calculate the amount of potential deposition and re-erosion at each time-step, then sum these to get an estimate of deposition over the whole tide. This procedure will be carried out for both the spring and the neap tide. A choice of half hourly time-steps is preferable to hourly time-steps, which would match the timing of the concentration measurements, because the duration of slack water periods may be quite short, and a long time-step might not resolve these properly. The hourly concentration measurements can be interpolated to give half-hourly values.

The measured concentrations are given in Table 14.

For each of the time-steps, we will estimate the near-bed concentration, by using the measured mid-depth concentration and hydrodynamic parameters to calculate the concentration profile. We will calculate the bed shear-stress from the water depth and the current speed, then we will calculate the rate of deposition or erosion.

In the absence of more accurate information from measurements or mathematical model results, assume that the variation of both the water level and the current speed are sinusoidal, and a quarter cycle out of phase, so that slack water occurs at high and low water, and peak current speeds occur at mid-tide level. Assume that the tidal period is 12·5 hours.

To illustrate the procedure, we will work through the details of the calculations for the time HW + 2 hours on the spring tide. At this time, the measured sediment concentration was $0{\cdot}17\,\mathrm{kg\,m^{-3}}$. Based on the sinusoidal curves (working in radians), the water depth in metres would be:

$$h = 5 + 1{\cdot}0 \times \cos\left(\frac{2\pi \times 2}{12{\cdot}5}\right) = 5{\cdot}54\,\mathrm{m}$$

*Table 14. Measured suspended-sediment concentrations at mid-depth, spring and neap tides*

| Time | Spring tide concentration: $kg\,m^{-3}$ | Neap tide concentration: $kg\,m^{-3}$ |
|---|---|---|
| HW | 0·15 | 0·07 |
| HW + 1 hour | 0·10 | 0·06 |
| HW + 2 | 0·17 | 0·09 |
| HW + 3 | 0·20 | 0·11 |
| HW + 4 | 0·26 | 0·14 |
| HW + 5 | 0·25 | 0·14 |
| HW + 6 | 0·18 | 0·07 |
| HW + 7 | 0·08 | 0·03 |
| HW + 8 | 0·14 | 0·03 |
| HW + 9 | 0·31 | 0·10 |
| HW + 10 | 0·28 | 0·12 |
| HW + 11 | 0·22 | 0·12 |
| HW + 12 | 0·10 | 0·07 |

over the shoal area and hence 10·54 m over the channel. The current speed over the shoal would be:

$$u = 1{\cdot}0 \times \sin\left(\frac{2\pi \times 2}{12{\cdot}5}\right) = 0{\cdot}84\,\mathrm{m\,s^{-1}}$$

Based on the conservation of water volume, the depth-averaged velocity in the channel area will be $0{\cdot}84 \times 5{\cdot}54/10{\cdot}54 = 0{\cdot}44\,\mathrm{m\,s^{-1}}$.

The near-bed concentration will be higher than that at mid-depth. Using the empirical concentration profile Equation (5.17), with the default value of $R_c = 3$ for the ratio of near-bed to near-surface concentration, allows the concentration at mid-depth to be related to the near-bed value. As this assumes a linear concentration profile, the mid-depth value will be the average of the near-surface and near-bed values. However, the near-surface value $C_{surf} = C_{bed}/3$. Therefore:

$$C_{mid} = \tfrac{1}{2}(C_{bed} + C_{bed}/3) = \tfrac{2}{3}C_{bed}$$

Therefore the near-bed concentration is 1·5 times larger than the mid-depth concentration. So we can multiply the measured value of $0{\cdot}17\,kg\,m^{-3}$ by this factor to obtain $0{\cdot}255\,kg\,m^{-3}$. When the flow crosses the channel, the detailed behaviour may be complex, while the concentration profile dynamically adjusts to a new situation, but for simplicity we will assume that the same near-bed profile exists in the deeper water of the channel. Note that because of the slow settling velocity of mud, it can take some considerable time to adjust to changes in the hydrodynamic conditions.

To calculate the bed shear-stress, we used the Christofferson and Jonsson (1985) method – see Equation (3.7) – implemented in Sandcalc–Hydrodynamics–Currents–Skin Friction–Smooth or Rough (Section 1.2). This method is valid for both rough turbulent and smooth turbulent situations. This method makes use of the median grain-size to determine the surface roughness. In the case of mud depositing in the channel, the surface of the bed is likely to be free of significant bed forms or roughness elements. When the grain size is small enough for the flow to be smooth turbulent, then the shear stress becomes independent of the grain size, so the exact value is not very important. A value of 50 microns was taken for the calculations presented below.

The critical bed shear-stress for deposition is normally assumed to be $0{\cdot}08\,N\,m^{-2}$, therefore at this current speed we would anticipate no deposition in the channel. The critical bed shear-stress for erosion is more variable and very difficult to predict accurately. If we assume that the channel will be a generally depositional environment, then it is likely to contain underconsolidated mud, relatively easy to erode. Therefore we will assume a critical shear-stress for erosion of $0{\cdot}2\,N\,m^{-2}$. The rate of erosion can be calculated using Equation 4.3, with the erosion rate constant taken to be $0{\cdot}002\,m^{-1}\,s$.

The calculation of bed shear-stress was carried out for each half hour time-step for both the spring and neap tides. On the spring tide, the bed shear-stress in the channel falls below the critical value for deposition around slack water, and exceeds

*Table 15. Calculation of deposition and erosion rate through the spring tide*

| Hours after HW | Conc. mid-depth: kg m$^{-3}$ | Conc. near bed: kg m$^{-3}$ | Depth: m | Depth in channel: m | Current speed: m s$^{-1}$ | Current speed in channel: m s$^{-1}$ | Bed shear-stress in channel: N m$^{-2}$ | Deposition (+ve) or erosion (−ve) per half hour: kg m$^{-2}$ per half hour |
|---|---|---|---|---|---|---|---|---|
| 0 | 0·15 | 0·225 | 6 | 11·00 | 0·00 | 0·00 | 0 | 0·4050 |
| 0·5 | 0·125 | 0·1875 | 5·97 | 10·97 | 0·25 | 0·12 | 0·018 | 0·2616 |
| 1 | 0·1 | 0·15 | 5·88 | 10·88 | 0·48 | 0·24 | 0·0632 | 0·0567 |
| 1·5 | 0·135 | 0·2025 | 5·73 | 10·73 | 0·68 | 0·34 | 0·1207 | 0·0000 |
| 2 | 0·17 | 0·255 | 5·54 | 10·54 | 0·84 | 0·42 | 0·1792 | 0·0000 |
| 2·5 | 0·185 | 0·2775 | 5·31 | 10·31 | 0·95 | 0·48 | 0·2308 | −0·1109 |
| 3 | 0·2 | 0·3 | 5·06 | 10·06 | 1·00 | 0·50 | 0·2495 | −0·1782 |
| 3·5 | 0·23 | 0·345 | 4·81 | 9·81 | 0·98 | 0·49 | 0·2412 | −0·1483 |
| 4 | 0·26 | 0·39 | 4·57 | 9·57 | 0·90 | 0·45 | 0·2066 | −0·0238 |
| 4·5 | 0·255 | 0·3825 | 4·36 | 9·36 | 0·77 | 0·39 | 0·1589 | 0·0000 |
| 5 | 0·25 | 0·375 | 4·19 | 9·19 | 0·59 | 0·29 | 0·092 | 0·0000 |
| 5·5 | 0·215 | 0·3225 | 4·07 | 9·07 | 0·37 | 0·18 | 0·0382 | 0·3033 |

| | | | | | | | | |
|---|---|---|---|---|---|---|---|---|
| 6 | 0·18 | 0·27 | 4·01 | 9·01 | 0·13 | 0·06 | 0·0051 | 0·4550 |
| 6·5 | 0·13 | 0·195 | 4·01 | 9·01 | −0·13 | −0·06 | 0·0051 | 0·3286 |
| 7 | 0·08 | 0·12 | 4·07 | 9·07 | −0·37 | −0·18 | 0·0382 | 0·1129 |
| 7·5 | 0·11 | 0·165 | 4·19 | 9·19 | −0·59 | −0·29 | 0·092 | 0·0000 |
| 8 | 0·14 | 0·21 | 4·36 | 9·36 | −0·77 | −0·39 | 0·1589 | 0·0000 |
| 8·5 | 0·225 | 0·3375 | 4·57 | 9·57 | −0·90 | −0·45 | 0·2066 | −0·0238 |
| 9 | 0·31 | 0·465 | 4·81 | 9·81 | −0·98 | −0·49 | 0·2412 | −0·1483 |
| 9·5 | 0·295 | 0·4425 | 5·06 | 10·06 | −1·00 | −0·50 | 0·2495 | −0·1782 |
| 10 | 0·28 | 0·42 | 5·31 | 10·31 | −0·95 | −0·48 | 0·2308 | −0·1109 |
| 10·5 | 0·25 | 0·375 | 5·54 | 10·54 | −0·84 | −0·42 | 0·1792 | 0·0000 |
| 11 | 0·22 | 0·33 | 5·73 | 10·73 | −0·68 | −0·34 | 0·1207 | 0·0000 |
| 11·5 | 0·16 | 0·24 | 5·88 | 10·88 | −0·48 | −0·24 | 0·0632 | 0·0907 |
| 12 | 0·1 | 0·15 | 5·97 | 10·97 | −0·25 | −0·12 | 0·018 | 0·2093 |
| Total deposition over the tide (kg m$^{-2}$/tide) | | | | | | | | 1·3007 |

*Table 16. Calculation of deposition and erosion rate through the neap tide*

| Hours after HW | Conc. mid-depth: $kg\,m^{-3}$ | Conc. near bed: $kg\,m^{-3}$ | Depth: m | Depth in channel: m | Current speed: $m\,s^{-1}$ | Current speed in channel: $m\,s^{-1}$ | Bed shear-stress in channel: $N\,m^{-2}$ | Deposition (+ve) or erosion (−ve) per half hour: $kg\,m^{-2}$ per half hour |
|---|---|---|---|---|---|---|---|---|
| 0 | 0·07 | 0·105 | 5·50 | 10·50 | 0·00 | 0·00 | 0 | 0·1890 |
| 0·5 | 0·065 | 0·0975 | 5·48 | 10·48 | 0·15 | 0·07 | 0·0075 | 0·1590 |
| 1 | 0·06 | 0·09 | 5·44 | 10·44 | 0·29 | 0·14 | 0·0235 | 0·1144 |
| 1·5 | 0·075 | 0·1125 | 5·36 | 10·36 | 0·41 | 0·21 | 0·0497 | 0·0767 |
| 2 | 0·09 | 0·135 | 5·27 | 10·27 | 0·51 | 0·25 | 0·0688 | 0·0340 |
| 2·5 | 0·1 | 0·15 | 5·15 | 10·15 | 0·57 | 0·29 | 0·0907 | 0·0000 |
| 3 | 0·11 | 0·165 | 5·03 | 10·03 | 0·60 | 0·30 | 0·0967 | 0·0000 |
| 3·5 | 0·125 | 0·1875 | 4·91 | 9·91 | 0·59 | 0·29 | 0·091 | 0·0000 |
| 4 | 0·14 | 0·21 | 4·79 | 9·79 | 0·54 | 0·27 | 0·08 | 0·0000 |
| 4·5 | 0·14 | 0·21 | 4·68 | 9·68 | 0·46 | 0·23 | 0·0595 | 0·0969 |
| 5 | 0·14 | 0·21 | 4·60 | 9·60 | 0·35 | 0·18 | 0·0379 | 0·1989 |
| 5·5 | 0·105 | 0·1575 | 4·54 | 9·54 | 0·22 | 0·11 | 0·0153 | 0·2293 |

| | | | | | | | | |
|---|---|---|---|---|---|---|---|---|
| 6 | 0·07 | 0·105 | 4·50 | 9·50 | 0·08 | 0·04 | 0·0024 | 0·1833 |
| 6·5 | 0·05 | 0·075 | 4·50 | 9·50 | −0·08 | −0·04 | 0·0024 | 0·1310 |
| 7 | 0·03 | 0·045 | 4·54 | 9·54 | −0·22 | −0·11 | 0·0153 | 0·0655 |
| 7·5 | 0·03 | 0·045 | 4·60 | 9·60 | −0·35 | −0·18 | 0·0379 | 0·0426 |
| 8 | 0·03 | 0·045 | 4·68 | 9·68 | −0·46 | −0·23 | 0·0595 | 0·0208 |
| 8·5 | 0·065 | 0·0975 | 4·79 | 9·79 | −0·54 | −0·27 | 0·08 | 0·0000 |
| 9 | 0·1 | 0·15 | 4·91 | 9·91 | −0·59 | −0·29 | 0·091 | 0·0000 |
| 9·5 | 0·11 | 0·165 | 5·03 | 10·03 | −0·60 | −0·30 | 0·0967 | 0·0000 |
| 10 | 0·12 | 0·18 | 5·15 | 10·15 | −0·57 | −0·29 | 0·0907 | 0·0000 |
| 10·5 | 0·12 | 0·18 | 5·27 | 10·27 | −0·51 | −0·25 | 0·0688 | 0·0454 |
| 11 | 0·12 | 0·18 | 5·36 | 10·36 | −0·41 | −0·21 | 0·0497 | 0·1227 |
| 11·5 | 0·095 | 0·1425 | 5·44 | 10·44 | −0·29 | −0·14 | 0·0235 | 0·1812 |
| 12 | 0·07 | 0·105 | 5·48 | 10·48 | −0·15 | −0·07 | 0·0075 | 0·1713 |
| Total deposition over the tide (kg $m^{-2}$/tide) | | | | | | | | 2·0619 |

the critical stress for erosion for a short period during the main run of the tide, so the net accretion in the channel will arise from a combination of both erosion and deposition. On the neap tide, the shear-stress does not reach the critical shear-stress for erosion, so only deposition takes place. We can then calculate the settling flux using Equation (8.2) and the erosion flux using Equation (4.3).

We will assume that the settling velocity has a constant value of $1\,\mathrm{mm\,s^{-1}}$. The settling flux is then calculated according to Equation (8.2), using the near-bed suspended-sediment concentrations calculated as described above. The results are presented in Table 15.

The total deposition in the channel over the spring tide is $1{\cdot}30\,\mathrm{kg\,m^{-2}}$. Repeating the calculation for a neap tide, with the neap tide measured concentrations, gives the results shown in Table 16.

Thus the net accretion over a neap tide is estimated to be $2{\cdot}06\,\mathrm{kg\,m^{-2}}$.

To estimate the annual siltation, we can approximate the whole year of varying conditions as 350 spring tides and 350 neap tides. This yields a total of $1176\,\mathrm{kg\,m^{-2}\,yr^{-1}}$. A typical bulk density of bed deposits of this type would be $1400\,\mathrm{kg\,m^{-3}}$. However, this figure also includes the water in the mud, whereas our deposition calculation is in terms of mass of dry sediment, therefore we must convert the density of bed deposits to a dry density. Using Table 2 to convert from bulk density ($\rho_B$) to dry density ($C_M$), assuming a sediment grain-density of $2650\,\mathrm{kg\,m^{-3}}$ and water density of $1027\,\mathrm{kg\,m^{-3}}$, gives $C_M = 609\,\mathrm{kg\,m^{-3}}$. Therefore the calculated deposition corresponds to a depth of approximately $1{\cdot}93\,\mathrm{m\,yr^{-1}}$.

## 13.4. SEDIMENT DISPERSION FROM A DREDGE DISPOSAL SITE

A potential site is being considered for disposal of muddy dredged material arising from maintenance of a port approach channel. It is desired that the sediment placed at the site should disperse, rather than accumulate. A typical dredging campaign requires disposal of $100\,000\,\mathrm{m^3}$ of material, placed at a density of $1300\,\mathrm{kg\,m^{-3}}$. All of the material is fine-grained, i.e. less than

63 microns. We will assume that the material is not dispersed during its descent to the bed at the placement site and that the material spreads to a layer of approximate thickness 0·5 m, covering a circular area of diameter 500 m. How long will the material take to disperse from the bed?

The key issue is the strength of currents and waves at the site. We will take a probabilistic approach, following the outline given in Soulsby (1997), Chapter 12 'Handling the wave-current climate'. Using this, the expected erosion rate can be calculated. We will neglect the possibility of sediment redepositing at the disposal site. This process, if it occurs, is likely to be very much slower than the erosion of the soft freshly-placed dredged mud.

The ideal method of establishing the climate of waves and currents at the site is by measuring them using a long-term deployment of a bed-mounted frame with current meter and pressure sensor. If that is not possible, they must be estimated from any available measurements and/or numerical model results.

The mean water depth at the site is 15 m. We will divide the range of possible current speeds at the site into 10 bands. The probability of the current falling into each band is given in Table 17.

*Table 17. Probability of occurrence for current speeds*

| Depth averaged current speed: $m\,s^{-1}$ | Probability of occurrence |
|---|---|
| 0·0–0·1 | 0·14 |
| 0·1–0·2 | 0·21 |
| 0·2–0·3 | 0·22 |
| 0·3–0·4 | 0·10 |
| 0·4–0·5 | 0·11 |
| 0·5–0·6 | 0·09 |
| 0·6–0·7 | 0·06 |
| 0·7–0·8 | 0·04 |
| 0·8–0·9 | 0·02 |
| 0·9–1·0 | 0·01 |

*Table 18. Probability of occurrence of wave orbital velocities*

| Wave orbital velocity: m s$^{-1}$ | Peak wave period: s | Probability of occurrence |
|---|---|---|
| 0·0–0·2 | 4·1 | 0·25 |
| 0·2–0·4 | 4·5 | 0·38 |
| 0·4–0·6 | 5·0 | 0·14 |
| 0·6–0·8 | 5·4 | 0·08 |
| 0·8–1·0 | 6·0 | 0·06 |
| 1·0–1·2 | 6·3 | 0·04 |
| 1·2–1·4 | 6·5 | 0·02 |
| 1·4–1·6 | 6·8 | 0·02 |
| 1·6–1·8 | 7·2 | 0·01 |
| 1·8–2·0 | 7·4 | 0·00 |

The same approach can be taken with wave-orbital velocity, which can be calculated from the wave height, period and water depth (Soulsby, 1997). The results are shown in Table 18.

This gives 100 combinations of wave and current condition. For each of these we require the joint probability and the combined wave-current bed shear-stress.

Firstly, the probabilities can be worked out using the usually accurate assumption that the current speed and the wave conditions are independent, by multiplying the probabilities for the separate wave and current conditions – see Table 19. This is only valid in water deep enough that tidal currents dominate over wave-induced currents.

For each combination of current speed and wave height, we can work out the bed shear-stress. Because of the fluid properties of the mud during the deposition process, the surface of the mud is likely to be very smooth and we can use the smooth turbulent law to calculate the bed shear-stress due to currents (see Equation (3.6), or Sandcalc–Hydrodynamics–Currents–Skin Friction–Logarithmic Smooth) (Section 1.2). For wave induced shear-stress we will use Equation (3.18), with the friction factor given by Equation (3.21). The wave and current contributions to total shear-stress can then be combined using Equation (3.25)

*Table 19. Joint probability of each current/wave combination*

| Current: m s$^{-1}$ | Wave orbital velocity: m s$^{-1}$ | | | | | | | | | |
|---|---|---|---|---|---|---|---|---|---|---|
| | 0·1 | 0·3 | 0·5 | 0·7 | 0·9 | 1·1 | 1·3 | 1·5 | 1·7 | 1·9 |
| 0·05 | 0·0350 | 0·0532 | 0·0196 | 0·0112 | 0·0084 | 0·0056 | 0·0028 | 0·0028 | 0·0014 | 0 |
| 0·15 | 0·0525 | 0·0798 | 0·0294 | 0·0168 | 0·0126 | 0·0084 | 0·0042 | 0·0042 | 0·0021 | 0 |
| 0·25 | 0·0550 | 0·0836 | 0·0308 | 0·0176 | 0·0132 | 0·0088 | 0·0044 | 0·0044 | 0·0022 | 0 |
| 0·35 | 0·0250 | 0·0380 | 0·0140 | 0·0080 | 0·0060 | 0·0040 | 0·0020 | 0·0020 | 0·0010 | 0 |
| 0·45 | 0·0275 | 0·0418 | 0·0154 | 0·0088 | 0·0066 | 0·0044 | 0·0022 | 0·0022 | 0·0011 | 0 |
| 0·55 | 0·0225 | 0·0342 | 0·0126 | 0·0072 | 0·0054 | 0·0036 | 0·0018 | 0·0018 | 0·0009 | 0 |
| 0·65 | 0·0150 | 0·0228 | 0·0084 | 0·0048 | 0·0036 | 0·0024 | 0·0012 | 0·0012 | 0·0006 | 0 |
| 0·75 | 0·0100 | 0·0152 | 0·0056 | 0·0032 | 0·0024 | 0·0016 | 0·0008 | 0·0008 | 0·0004 | 0 |
| 0·85 | 0·0050 | 0·0076 | 0·0028 | 0·0016 | 0·0012 | 0·0008 | 0·0004 | 0·0004 | 0·0002 | 0 |
| 0·95 | 0·0025 | 0·0038 | 0·0014 | 0·0008 | 0·0006 | 0·0004 | 0·0002 | 0·0002 | 0·0001 | 0 |

to calculate the cycle-mean bed shear-stress. However, the most relevant parameter for erosion is the maximum shear-stress over the cycle. This can be obtained using Equation (3.26). We will assume that the angle between the waves and the currents is 90°, typical of a coastal situation with shore parallel currents and shore normal waves.

This leads to the results shown in Table 20 of maximum combined wave and current shear-stresses.

For each value of combined maximum wave and current shear-stress, we can evaluate the erosion rate, using Equation (4.3) or Equation (4.4), with a value of $m_e = 0{\cdot}002\,\mathrm{m}^{-1}\,\mathrm{s}$, as indicated in Figure 16. In the erosion rate equation, we use the maximum combined wave and current shear-stress, replacing the current shear-stress from Equation (4.3) or the wave shear-stress from Equation (4.4).

In this example, we have assumed that the mechanism of erosion by waves is essentially the same as that due to currents and so an enhanced shear-stress has been calculated and applied to Equation (4.3). As discussed in Sections 4.3 and 6.3, in some circumstances wave action can fluidise the bed. In that case, the rate of erosion is not so directly related to the bed shear-stress associated with the wave and depends in a complex way on the properties and structure of the bed.

Measurements of properties of dredged material (Feates and Mitchener, 1998) indicate that the critical shear-stress for erosion should be approximately $0{\cdot}2\,\mathrm{N\,m}^{-2}$. We then have a table of 100 values of erosion rate, associated with each combination of waves and currents. The calculated dry mass erosion rates are as shown in Table 21.

To get the expected average erosion rate, we must multiply each erosion rate by the associated probability and add all of the values together.

Note that because of the threshold in the calculation of erosion, we have taken the approach of calculating the average erosion rate. This is not the same in general as the erosion rate corresponding to the average shear-stress, as can be seen by considering an example of an average shear-stress slightly less than threshold, resulting from periods of stress both higher and lower than the threshold. The high stress periods will result in erosion, whereas if only the average stress were considered, because it lies below the threshold, the estimated erosion rate would be zero.

*Table 20. The maximum combined wave current shear-stresses (units are in $N\,m^{-2}$)*

| Current: $m\,s^{-1}$ | Wave orbital velocity: $m\,s^{-1}$ | | | | | | | | | |
|---|---|---|---|---|---|---|---|---|---|---|
| | 0·1 | 0·3 | 0·5 | 0·7 | 0·9 | 1·1 | 1·3 | 1·5 | 1·7 | 1·9 |
| 0·05 | 0·1508 | 0·4257 | 0·6720 | 0·9049 | 1·8187 | 2·4973 | 3·2575 | 4·0763 | 4·9432 | 5·8928 |
| 0·15 | 0·1687 | 0·4538 | 0·6971 | 0·9267 | 1·8324 | 2·5079 | 3·2660 | 4·0832 | 4·9490 | 5·8978 |
| 0·25 | 0·1720 | 0·4816 | 0·7434 | 0·9801 | 1·8835 | 2·5524 | 3·3041 | 4·1161 | 4·9775 | 5·9226 |
| 0·35 | 0·1947 | 0·4853 | 0·7646 | 1·0205 | 1·9599 | 2·6323 | 3·3810 | 4·1874 | 5·0428 | 5·9818 |
| 0·45 | 0·2443 | 0·4941 | 0·7673 | 1·0322 | 2·0263 | 2·7216 | 3·4813 | 4·2906 | 5·1443 | 6·0791 |
| 0·55 | 0·3175 | 0·5238 | 0·7761 | 1·0345 | 2·0632 | 2·7921 | 3·5783 | 4·4042 | 5·2665 | 6·2045 |
| 0·65 | 0·4106 | 0·5789 | 0·8039 | 1·0465 | 2·0753 | 2·8326 | 3·6516 | 4·5053 | 5·3879 | 6·3398 |
| 0·75 | 0·5213 | 0·6589 | 0·8552 | 1·0774 | 2·0779 | 2·8485 | 3·6951 | 4·5800 | 5·4907 | 6·4660 |
| 0·85 | 0·6483 | 0·7619 | 0·9314 | 1·1312 | 2·0850 | 2·8524 | 3·7143 | 4·6253 | 5·5654 | 6·5696 |
| 0·95 | 0·7905 | 0·8855 | 1·0317 | 1·2093 | 2·1055 | 2·8567 | 3·7200 | 4·6470 | 5·6115 | 6·6446 |

*Table 21. Dry mass erosion rates for the combined wave-current conditions in Table 20 (units are kg m$^{-2}$ s$^{-1}$)*

| Current: m s$^{-1}$ | Wave orbital velocity: m s$^{-1}$ | | | | | | | | | |
|---|---|---|---|---|---|---|---|---|---|---|
| | 0·1 | 0·3 | 0·5 | 0·7 | 0·9 | 1·1 | 1·3 | 1·5 | 1·7 | 1·9 |
| 0·05 | 0·0E + 00 | 4·5E – 04 | 9·4E – 04 | 1·4E – 03 | 3·2E – 03 | 4·6E – 03 | 6·1E – 03 | 7·8E – 03 | 9·5E – 03 | 1·1E – 02 |
| 0·15 | 0·0E + 00 | 5·1E – 04 | 9·9E – 04 | 1·5E – 03 | 3·3E – 03 | 4·6E – 03 | 6·1E – 03 | 7·8E – 03 | 9·5E – 03 | 1·1E – 02 |
| 0·25 | 0·0E + 00 | 5·6E – 04 | 1·1E – 03 | 1·6E – 03 | 3·4E – 03 | 4·7E – 03 | 6·2E – 03 | 7·8E – 03 | 9·6E – 03 | 1·1E – 02 |
| 0·35 | 0·0E + 00 | 5·7E – 04 | 1·1E – 03 | 1·6E – 03 | 3·5E – 03 | 4·9E – 03 | 6·4E – 03 | 8·0E – 03 | 9·7E – 03 | 1·2E – 02 |
| 0·45 | 8·9E – 05 | 5·9E – 04 | 1·1E – 03 | 1·7E – 03 | 3·7E – 03 | 5·0E – 03 | 6·6E – 03 | 8·2E – 03 | 9·9E – 03 | 1·2E – 02 |
| 0·55 | 2·4E – 04 | 6·5E – 04 | 1·2E – 03 | 1·7E – 03 | 3·7E – 03 | 5·2E – 03 | 6·8E – 03 | 8·4E – 03 | 1·0E – 02 | 1·2E – 02 |
| 0·65 | 4·2E – 04 | 7·6E – 04 | 1·2E – 03 | 1·7E – 03 | 3·8E – 03 | 5·3E – 03 | 6·9E – 03 | 8·6E – 03 | 1·0E – 02 | 1·2E – 02 |
| 0·75 | 6·4E – 04 | 9·2E – 04 | 1·3E – 03 | 1·8E – 03 | 3·8E – 03 | 5·3E – 03 | 7·0E – 03 | 8·8E – 03 | 1·1E – 02 | 1·3E – 02 |
| 0·85 | 9·0E – 04 | 1·1E – 03 | 1·5E – 03 | 1·9E – 03 | 3·8E – 03 | 5·3E – 03 | 7·0E – 03 | 8·9E – 03 | 1·1E – 02 | 1·3E – 02 |
| 0·95 | 1·2E – 03 | 1·4E – 03 | 1·7E – 03 | 2·0E – 03 | 3·8E – 03 | 5·3E – 03 | 7·0E – 03 | 8·9E – 03 | 1·1E – 02 | 1·3E – 02 |

The value obtained for the average erosion rate is $0{\cdot}00132\,\mathrm{kg\,m^{-2}\,s^{-1}}$, equivalent to $114\,\mathrm{kg\,m^{-2}\,day^{-1}}$. To relate this to the depth of the sediment deposit, we must work out a dry density of the deposit from the given value of bulk density, $\rho_B = 1300\,\mathrm{kg\,m^{-3}}$. Using Table 2, assuming a grain density $\rho_s = 2650\,\mathrm{kg\,m^{-3}}$ and a water density of $1027\,\mathrm{kg\,m^{-3}}$ gives $C_M = 446\,\mathrm{kg\,m^{-3}}$. Because the initial deposit was 0·5 m thick, at the average erosion rate, this will take just under two days to erode all of the material disposed of at the site. Therefore, we can be confident that the sediment disposed of at the site will be dispersed quickly and will not accumulate.

Note that large wave events have a high erosion rate, but a low joint probability of occurrence, whereas mild wave conditions have a low erosion rate, but occur frequently. Based on Tables 19 and 21 divided into 3 blocks of 3 columns each, representing, mild (0·1, 0·3, $0{\cdot}5\,\mathrm{m\,s^{-1}}$), moderate (0·7, 0·9, $1{\cdot}1\,\mathrm{m\,s^{-1}}$) and severe (1·3, 1·5, $1{\cdot}7\,\mathrm{m\,s^{-1}}$) wave conditions, we see that 31% of the erosion occurs during mild waves (77% of the time), 40% during moderate waves (18% of the time) and 29% during severe waves (5% of the time). This division is arbitrary, but it can be seen that all parts of the wave height spectrum have, very approximately, an equal influence in this case. In terms of the time taken to erode the deposited sediments, this means that while the average time for erosion is two days, there may be a few occasions when it erodes much more quickly owing to a severe storm, offset by many occasions when it erodes more slowly than the average. If only mild waves occur, then the erosion rate, normalised to adjust from 77% probability to 100% probability, becomes $45{\cdot}5\,\mathrm{kg\,m^{-2}\,day^{-1}}$ and so the time taken to erode all of the material becomes 4·9 days.

The case studies presented are intended to indicate representative methodologies and use typical input values to demonstrate some of the approaches contained in the manual. The final answers obtained in the calculations are not intended to form guidance and new calculations will need to be made for a specific study. In practice the methodology for a particular study will also vary to suit the scope of that study, i.e. more or less complex, or even different, calculations will be required.

# References

Alvarez-Hernandez, E. (1990). *The influence of cohesion on sediment movement in channels of circular cross section.* PhD thesis, University of Newcastle-upon-Tyne.

Amos, C. L., Grant, J., Daborn, G. R. and Black, K. (1992). Sea Carousel — a benthic, annular flume. *Estuarine, Coastal and Shelf Science*, **34**, 557–577.

Amos, C. L., Brylinsky, M., Sutherland, T. F., O'Brien, D., Lee, S. and Cramp, A. (1998). The stability of a mudflat in the Humber Estuary, South Yorkshire, UK. In: *Sedimentary processes in the intertidal zone*, eds, K. S. Black, D. M. Paterson and A. Cramp. Geological Society, London, Special Publication 139, 25–43.

Arnskov, M. M., Fredsøe, J. and Sumer, B. M. (1993). Bed shear stress measurements over a smooth bed in three-dimensional wave-current motion. *Coastal Engng*, **20**, 277–316.

Bale, A. J. and Morris, A. W. (1987). In-situ measurements of particle size in estuarine waters. *Estuarine, Coastal and Shelf Science*, **24**, 253–263.

Barbier, J. M. (1977). Study of the mechanism of mud movement in the Gironde. *Proc. 24th Int. Navigation Congress*. Leningrad.

Batchelor, G. K. and Green, J. T. (1972). The determination of the bulk stress in a suspension of spherical particles to order $c^2$. *J. Fluid Mech.*, **56** (3), 401–427.

Beeftink, W. G., Nieuwenhuize, J., Stoeppler, M. and Mohl, C. (1982). Heavy metal accumulation in saltmarshes from the western and eastern Sheldt. *Science of the Total Environment*, **26**, 255–312.

Berlamont, J., Ockenden, M. C., Toorman, E. and Winterwerp, J. (1993). The characterisation of cohesive sediment properties. *Coastal Engng*, **21**, 105–128.

Black, K. S. and Paterson, D. M. (1997). Measurement of the erosion potential of cohesive marine sediments: a review of current in-situ technology. *J. Marine Env. Engng*, **4**, 43–83.

Black, K. S., Paterson, D. M. and Cramp, A., eds. (1998). *Sedimentary processes in the intertidal zone*. Geological Society, London, Special Publication 139.

Bryant, R., James, A. E. and Williams, D. J. A. (1980). Rheology of cohesive sediments. In: *Industrialised embayments and their environmental problems. A case study of Swansea Bay*. eds, M. B. Collins, F. T. Banner, P. A. Tyler, S. J. Wakefield and A. E. James. Pergamon Press.

Burt, T. N. and Game, A. C. (1985). *Deposition of fine sediment from flowing water: an investigation of dependence on concentration*. HR Wallingford Report SR 27.

Burt, T. N. and Stevenson, J. R. (1983). *Field settling velocity of Thames mud*. HR Wallingford Report IT 251.

Chesher, T. J. and Ockenden, M. C. (1994). *Tidal transport of mud/sand mixtures. Development of an integrated numerical model*. HR Wallingford Report SR 380.

Christofferson, J. B. and Jonsson, I. G. (1985). Bed friction and dissipation in a combined current and wave motion. *Ocean Engng*, **12**, 387–423.

Cooper, A. J. and Dearnaley, M. P. (1996). *Guidelines for the use of computational models in coastal and estuarial studies. Flow and sediment transport models*. HR Wallingford Report SR 456.

Coussot, P. (1997). Mudflow rheology and dynamics. Balkema, Rotterdam.

Dade, B. W., Davies, J. D., Nichols, P. D., Nowell, A. R. M., Thistle, D., Trexler, M. B. and White, D. C. (1990). Effects of bacterial exopolymer adhesion on the entrainment of marine sand. *Geomicrobiology J.*, **8**, 1–16.

Dade, B. W. and Nowell, A. R. M. (1991). Moving muds in the marine environment. *Proc. Coastal Sediments '91 Conf.*, 54–71, ASCE.

Dearnaley, M. P. (1991). *Flocculation and settling of cohesive sediments*. HR Wallingford Report SR 272.

Dearnaley, M. P. (1997). Direct measurements of settling velocities in the Owen Tube: a comparison with gravimetric analysis. In: *Cohesive Sediments*, eds, N. Burt, R. Parker and J. Watts, pp. 75–85. Wiley, Chichester.

Dearnaley, M. P., Roberts, W., Spearman, J. and Cooper, A. J. (1997). *Development of a unified approach for assessing sedimentation in harbours*. HR Wallingford Report SR 497.

Deigaard, R., Bro Mikkelsen, M. and Fredsøe, J. (1991). *Measurements of the bed shear stress in a surf zone*. Progress Report 73. Inst. Hydrodynamic and Hydraulic Engng, Technical University of Denmark, Lyngby, 21–30.

Delo, E. A. (1987). *The behaviour of estuarine muds during tidal cycles. An experimental study and mathematical model.* HR Wallingford Report SR 138.

Delo, E. A. (1988a). *Estuarine muds manual.* HR Wallingford Report SR 164.

Delo, E. A. (1988b). *Deposition of cohesive sediment from flowing water.* HR Wallingford Report SR 152.

Delo, E. A. and Burt, T. N. (1986). *The hydraulic engineering characteristics of estuarine muds. A working manual and literature review.* HR Wallingford Report SR 77.

Delo, E. A. and Ockenden, M. C. (1989). *Prediction of siltation at a point.* Paper presented at The Third National Conf. Hydraulic Engng, New Orleans, USA, 14–18, August 1989.

Delo, E. A. and Ockenden, M. C. (1992). *Estuarine muds manual.* HR Wallingford Report SR 309.

Derbyshire, B. V. and Kendrick, M. (1987). *Laboratory experiments on a near-bed turbid layer.* HR Wallingford Report SR 88.

Diserens, A. P. and Delo, E. A. (1988). *The stability of cohesive dredged slopes under wave action.* HR Wallingford Report SR 154.

Diserens, A. P., Delo, E. A. and Ockenden, M. C. (1991). *Estuarine sediments — near-bed processes: field measurement of near-bed cohesive sediment processes.* HR Wallingford Report SR 262.

Dyer, K. R. (1986). *Coastal and estuarine sediment dynamics.* John Wiley & Sons.

Dyer, K. R. (1998a). The typology of intertidal mudflats. In: *Sedimentary processes in the intertidal zone*, eds, K. S. Black, D. M. Paterson and A. Cramp. Special Publications 139, 11–24. Geological Society, London.

Dyer, K. R. (1998b). *Estuaries — a physical introduction.* 2nd Edition. Wiley, Chichester.

Dyer, K. R. (Guest ed.) (2000). Intertidal mudflats: properties and processes. Special issue of nearshore and coastal oceanography, eds, D. A. Huntley and J. Oltman-Shay. *Continental shelf research*, **20**, (Part I and Part II). Pergamon Press.

Dyer, K. R., Christie, M. C. and Wright, E. W. (2000). The classification of intertidal mudlats. *Continental shelf research*, **20**, 1039–1060. Pergamon Press.

Dyer, K. R., Cornelisse, J., Dearnaley, M. P., Fennessy, M. J., Jones, S. E., Kappenberg, J., McCave, I. N., Pejrup, M., Puls, W., Van Leussen, W. and Wolfstein, K. (1996). A comparison of in-situ techniques for estuarine floc settling velocity measurements. *J. Sea Res.*, **36**(1/2), 15–29.

Eckman, J. E., Nowell, A. R. M. and Jumars, P. A. (1981). Sediment destabilisation by animal tubes. *J. Marine Res.*, **39**, 361–374.

Eisma, K., Dyer, K. R. and Van Leussen, W. (1997). The in-situ determination of the settling velocities of suspended fine grained sediment — a review. In: *Cohesive sediments*, eds, N. Burt, R. Parker and J. Watts. 17–44, John Wiley & Sons Ltd.

Feates, N. G. and Mitchener, H. J. (1996). *Erosion shear stress measurements on sea bed cores taken from Sellafield mud patch 26 May–31 July 1996.* HR Wallingford Report TR 14.

Feates, N. G. and Mitchener, H. J. (1998). *Properties of dredged material: measurement of sediment properties of dredged material from Harwich Harbour.* HR Wallingford Report TR 46.

Fennessy, M. J., Dyer, K. R. and Huntley, D. A. (1994). INSSEV: an instrument to measure the size and settling velocity of flocs in situ. *Marine Geology*, **117**, 107–117.

Fredsøe, J. (1984). Turbulent boundary layer in wave-current motion. *J. Hydraul. Engng*, **110**, 1103–1120. ASCE.

Frostick, L. E. and McCave, I. N. (1979). Seasonal shifts of sediment within an estuary mediated by algal growth. *Estuarine and Coastal Marine Science*, **9**, 569–576.

Ginsberg, R. N. and Lowenstam, H. A. (1958). The influence of bottom marine communities on the depositional environment of sediments. *J. Geology*, **66**, 310–318.

Grant, J. and Gust, G. (1987). Prediction of coastal sediment stability from phytopigment content of mats of purple sulphur bacteria. *Nature*, **9**, 244–246.

Grant, J. and Madsen, O. S. (1979). Combined wave and current interaction with a rough bottom. *J. Geophys. Res.*, **84**(C4), 1797–1808.

Harleman, D. (1961). Stratified flow. In: *Handbook of fluid dynamics*, ed. V. Streeter. McGraw-Hill, New York, USA.

Hayter, E. J. and Mehta, A. J., (1982). *Modelling of estuarial fine sediment transport for tracking pollutant movement.* Final Report No. UFL/COEL-8L/009. University of Florida.

Heinzelmann, C. H. and Wallisch, S. (1991). Benthic settlement and bed erosion. A Review. *J. Hyd. Res.*, **29**, 355–371.

HR Wallingford (1989). *Grangemouth mud properties.* HR Wallingford Report SR 197.

HR Wallingford (1990). *Fluid mud in estuaries. Laboratory experiments.* HR Wallingford Report EX2085.

HR Wallingford (1991). *Fluid mud in estuaries: final report to ETSU*. HR Wallingford Report EX2392.

Huxley, T. H. (1868). On some organisms living at great depths in the North Atlantic Ocean. *Quarterly J. Microscopy*, **VIII**, 203–212.

INTRMUD Protocols. (1997). Compiled by Sediment Ecology Research Group, Gatty Marine Laboratory, St Andrews, Scotland, February 1997. EC INTRMUD Project.

James, A. E. and Williams D. J. A. (1982). Flocculation and rheology of kaolinite/quartz suspensions. *Rheologica Acta*, **21**, 176–183.

James, A. E., Williams, D. J. A. and Williams, P. R. (1987). Direct measurement of static yield properties of cohesive suspensions. *Rheologica Acta*, **26**, 437–446.

Jones, T. E. R. and Golden, K. (1990). *Rheological behaviour of River Parrett fluid mud.* Report contained in HR Wallingford (1990).

Kamphuis, J. W. (1990). Influence of sand or gravel on the erosion of cohesive sediment. *J. Hyd. Res.*, **28**(1), 43–53.

Kamphuis, W. and Hall, K. R. (1983). Cohesive material erosion by unidirectional current. *J. Hyd. Engng*, **109**, 49–61. ASCE.

Kersten, M. K. (1988). Geobiological effects on the mobility of contaminants in marine sediments. In: *Pollution of the North Sea: an assessment*, eds, W. Salomons, B. L. Bayne, E. K. Duursma and U. Forstner. Springer-Verlag, New York.

Kirby, R. (1986). *Suspended fine cohesive sediment in the Severn Estuary and inner Bristol Channel, UK.* Report prepared by Ravensrodd Consultants Ltd for UKAEA. Report No. ETSU-STP-4042.

Kirby, R. and Parker, W. R. (1973). Fluid mud in the Severn Estuary and the Bristol Channel and its relevance to pollution studies. *Annual Symp. Estuarine & Coastal Pollution.* Inst. of Chem. Engrs, Exeter.

Kirby, R. and Parker, W. R. (1983). Distribution and behaviour of fine sediment in the Severn Estuary and inner Bristol Channel, UK. *Canadian J. Fisheries & Aquatic Science*, **40**(suppl. 1), 83–95.

Krishnappan, B. G. (1991). Modelling of cohesive sediment transport. *Preprints Int. Symp. Trans. Suspended Sediments & its Mathematical Modelling*, IAHR, Florence, Italy, September 2–5 1991, 433–448.

Krone, R. B. (1962). *Flume studies of the transport of sediment in estuarial shoaling processes, final report.* Hydraulic Engineering and Sanitary Engineering Research Laboratory, University of California, Berkeley (CA).

Krone, R. B. (1972). *A field study of flocculation as a factor in estuarial shoaling processes.* U.S Army Corps of Engineering, Comm. Tidal Hydraulics, Technical Bulletin, No. 19.

Lee, S.-C. and Mehta, A. J. (1997). Problems in characterizing dynamics of mud shore profiles. *J. Hyd. Engng*, **123**(4), 351–361. ASCE.

Li, M. Z. and Gust, G. (2000). Boundary layer dynamics and drag reduction in flows of high cohesive sediment suspensions. *Sedimentology*, **47**(1), 71–86.

Maa, P. T. and Mehta, A. J. (1985). *Considerations on bed response on wave resuspension of muds*. Paper Presented at the Symp. Circulation Patterns in Estuaries, Virginia Institute of Marine Science, Gloucester Point, Virginia.

MAST G6M (1993). *On the methodology and accuracy of measuring physico-chemical properties to characterize cohesive sediments*. MAST G6M Coastal Morphodynamics Report to the EC MAST-I research programme, Commission of the European Communities DGXII, July 1993. Published by Delft Hydraulics.

Meadows, P. S. and Campbell, J. I. (1978). *An introduction to marine science*. Blackie and Son Limited, London & Glasgow.

Meadows, P. S. and Meadows, A. (1991). The environmental impact of burrows and burrowing animals. *Proc. Symp. Zoological Soc. London*, Clarendon Press, Oxford.

Mehta, A. J. (1988). Laboratory studies on cohesive sediment deposition and erosion. In: *Physical processes in estuaries*, eds, J. Dronkers and W. Van Leussen. Springer-Verlag, New York.

Mehta, A. J. and Dyer, K. R. (1990). Cohesive sediment transport in estuarine and coastal waters. In: *The sea: ideas and observations on progress in the study of the seas*, Vol. 9, Part B, Ch. 23, eds, B. Le Méhauté and D. M. Hanes, 815–839. John Wiley & Sons.

Mehta, A. J. and Lott, J. W. (1987). Sorting of fine sediment during deposition. *Coastal sediments: Proc. Conf. Advances in Understanding of Coastal Sediment Processes*, New Orleans. Vol 1.

Mehta, A. J. and Partheniades, E. (1973). *Depositional behaviour of cohesive sediments*. Technical Report No. 16. Coastal and Oceanographic Engineering Laboratory, University of Florida, Gainesville (FL).

Migniot, C. (1968). A study of the physical properties of various very fine sediments and their behaviour under hydrodynamic action. *La Houille Blanche*, **23**(7). (Translation of French text.)

Mitchener, H. J. (1996). *Properties of dredged material. A review of available measurement techniques for determining physical properties*. HR Wallingford Report TR 21.

Mitchener, H. J., Torfs, H. and Whitehouse, R. J. S. (1996a). Erosion of mud/sand mixtures. *Coastal Engng*, **29**, 1–25. (Errata, 1997, **30**, 319.)

Mitchener, H. J., Whitehouse, R. J. S., Soulsby, R. L. and Lawford, V. A. (1996b). *Estuarine morphodynamics. Development of SedErode — instrument for in-situ mud erosion measurements.* HR Wallingford Report TR 17.

Myrhaug, D. and Slaattelid, O. H. (1990). A rational approach to wave-current friction coefficients for rough, smooth and transitional turbulent flow. *Coastal Engng*, **14**, 265–293.

O'Brien, D. J., Whitehouse, R. J. S. and Cramp, A. (2000). The cyclic development of a macrotidal mudflat on varying timescales. *Continental shelf research*, **20**, 1593–1619. Pergamon Press.

O'Brien, J. S. and Julien, P. Y. (1986). Rheology of non-Newtonian fine sediment mixtures. *Proc. ASCE Speciality Conf. on Aerodynamics, Fluid Mechanics and Hydraulics, Minneapolis, Minnesota*, 989–996.

O'Brien, J. S. and Julien, P. Y. (1988). Laboratory analysis of mudflow properties. *J. Hyd. Engng*, **114**(8), 877–887. ASCE.

O'Connor, B. A. and Tuxford, C. (1980). Modelling siltation at dock entrances. *Third Int. Symp. Dredging Tech.*, Bordeaux, 359–372, BHRA Cranfield.

Ockenden, M. C. (1993). A model for the settling of non-uniform cohesive sediment in a laboratory flume and an estuarine field setting. *J. Coastal Res.*, **9**(4), 1094–1105.

Ockenden, M. C. and Delo, E. A. (1991). Laboratory testing of muds. *Geo-Marine Letters*, **11**, 138–142.

Odd, N. V. M. and Cooper, A. J. (1989). A two-dimensional model of the movement of fluid mud in a high energy turbid estuary. *J. Coastal Res.*, **5** (special issue), 185–193.

Odd, N. V. M. and Owen, M. W. (1972). A two-layer model of mud transport in the Thames estuary. *Proc. ICE Supplement (ix)*. Paper 7517S, 175–205.

Odd, N. V. M. and Rodger, J. G. (1986). *An analysis of the behaviour of fluid mud in estuaries.* HR Wallingford Report SR 84.

Owen, M. W. (1971). The effect of turbulence on the settling velocities of silt flocs. *Proc. Int. Assoc. Hyd. Res., 14th Conf.*, **4**, 27–32, Paris.

Owen, M. W. (1977). Problems in the modelling of transport, erosion, and deposition of cohesive sediments. In: *The sea: ideas and observations on progress in the study of the seas*, Vol. 6, Ch. 12, ed., E. D. Goldberg, 515–537. John Wiley & Sons Ltd.

Panagiotopoulos, I., Voulgaris, G. and Collins, M. B. (1997). The influence of clay on the threshold of movement of fine sandy beds. *Coastal Engng*, **32**, 19–43.

Partheniades, E. (1962). *A study of erosion and deposition of cohesive sediment in salt water.* PhD Thesis, University of California.

Partheniades, E. (1965). Erosion and deposition of cohesive soils. *Proc. Am. Soc. Civ. Engrs*, **91** (HY1) 105–139.

Partheniades, E., Cross, R. H. and Ayora, A. (1968). Further results on the deposition of cohesive sediments. *Second Conf. Coastal Engng*, Vol. 1, Ch. 47, London, September 1968.

Partheniades, E., Kennedy, J. F., Etter, R. J. and Hoyer, R. P. (1966). *Investigations of the depositional behaviour of fine cohesive sediments in an annular rotating channel.* Hydrodynamics Laboratory Report No. 96.

Paterson, D. M. (1994). Biological mediation of sediment erodibility: ecology and physical dynamics. In: *Cohesive sediments*, eds, N. Burt, R. Parker and J. Watts, 215–229. John Wiley & Sons Ltd.

Paterson, D. M., Crawford, R. M. and Little, C. (1990). Subaerial exposure and changes in the stability of intertidal estuarine sediments. *Estuarine, Coastal and Shelf Science*, **30**, 541–556.

Postma, H. (1962). Sediment. In: *Demerara coastal investigation: Delft, Netherlands.* Hydrodynamics Laboratory, 105–164.

Puls, W. and Kuehl, H. (1986). Field measurements of the settling velocity of estuarine flocs. *Third Int. Symp. River Sedimentation*, Jackson, Mississippi, USA. March 1986.

Raudkivi, A. J. (1998). *Loose boundary hydraulics.* 3rd Edition. Balkema, Rotterdam.

Roberts, W. (1993). Development of a mathematical model of fluid mud in the coastal zone. *Proc. Instn Civ. Engrs Wat., Marit. & Energy*, **101**, Sept., 173–181.

Roberts, W., Le Hir, P. and Whitehouse, R. J. S. (2000). Investigation using simple mathematical models of the effect of tidal currents and waves on the profile shape of intertidal mudflats. *Continental shelf research*, **20**, 1079–1097. Pergamon Press.

Sauriau, P.-G., Orvain, F. and Le Hir, P. (1999). Bioturbation influence of the mud snail *Hydrobia ulvae* on cohesive sediment erodability: results from the application of a 1-DV model. *Intertidal mudflats: properties and processes.* University of Plymouth, Plymouth, England, 12–16 April 1999. Abstract volume.

Sills, G. C. (1997). Consolidation of cohesive sediments in settling columns. In: Cohesive Sediments, eds N. Burt, R. Parker and J. Watts, 107–120, John Wiley & Sons Ltd.

Sleath, J. F. A. (1990). Velocities and bed friction in combined flows. *Proc. 22nd Int. Conf. Coastal Engng*, Delft, Ch. 35, 450–463.

Salomons, W., Bayne, B. L., Duursma, E. K. and Forstner, U. (eds) (1988). *Pollution of the North Sea: an assessment.* Springer-Verlag, New York.

Soulsby, R. L. (1995). Bed shear-stresses due to combined waves and currents. In: *Advances in coastal morphodynamics*, eds, M. J. F. Stive, H. J. de Vriend, J. Fredsøe, L. Hamm, R. L. Soulsby, C. Teisson and J. C. Winterwerp, 4-20 to 4-23. Delft Hydraulics, The Netherlands.

Soulsby, R. L. (1997). *Dynamics of marine sands. A manual for practical applications*. Thomas Telford, London.

Soulsby, R. L. (2000). *Methods for predicting suspensions of mud*. HR Wallingford Report TR 104.

Soulsby, R. L., Hamm, L., Klopman, G., Myrhaug, D., Simons, R. R. and Thomas, G. P. (1993). Wave-current interaction within and outside the bottom boundary layer. *Coastal Engng*, **21**, 41–69.

Soulsby, R. L. and Smallman, J. V. (1986). *A direct method of calculating bottom orbital velocity under waves*. HR Wallingford Report SR 76.

Stevenson, J. R. and Burt, T. N. (1985). *Further studies of the field settling velocities of Thames mud*. HR Wallingford Report SR 10.

Stive, M. J. F., de Vriend, H. J., Fredsøe, J., Hamm, L., Soulsby, R. L., Teisson, C. and Winterwerp, J. C. (eds) (1995). *Advances in coastal morphodynamics*. Delft Hydraulics, Netherlands.

Teisson, C., Ockenden, M., Le Hir, P., Kranenburg, C. and Hamm, L. (1993). Cohesive sediment transport processes. *Coastal Engng*, **21**, 129–162.

Terzaghi, K., Peck, R. B. and Mesri, G. (1996). *Soil mechanics in engineering practice*. John Wiley & Sons.

Thorn, M. F. C. and Parsons J. G. (1980). Erosion of cohesive sediments in estuaries: an engineering guide. *3rd Int. Symp. Dredging Technology*, Bordeaux, 349–358. BHRA, Cranfield.

Torfs, H. (1994). Erosion of layered sand-mud beds in uniform flow. *Proc. 24th Int. Conf. Coastal Engng*, Kobe, Japan, 23–28, October 1994.

Torfs, H., Mitchener, H. J., Huysentruyt, H. and Toorman, E. (1996). Settling and consolidation of mud/sand mixtures. *Coastal Engng*, **29**, 27–45.

US Army Corps of Engineers (USACE) (1998). *Coastal Engineering Manual, Part III*. Circular No. 1110-2-292. Department of the Army, US Army Corps of Engineers, Washington DC.

Van der Lee, W. T. B. (1998). The impact of fluid shear and the suspended sediment concentration on the mud floc size variation in the Dollard Estuary, The Netherlands. In: *Sedimentary processes in the intertidal zone*, eds, K. S. Black, D. M. Paterson and A. Cramp. Special Publication 139, Geological Society, 187–198.

Van Kessel, T. and Blom, C. (1998). Rheology of cohesive sediments: comparison between a natural and an artificial mud. *J. Hyd. Res.*, **36**(4), 591–612.

Van Leussen, W. (1988). Aggregation of particles, settling velocity of mud flocs, a review. *Proc. Int. Symp. Physical Processes in Estuaries*, 347–403. Springer-Verlag, New York.

Van Leussen, W. (1997). The Kolmogorov microscale as a limiting value for the floc sizes of suspended fine-grained sediment in estuaries. In: *Cohesive Sediments*, eds, N. Burt, R. Parker and J. Watts. 45–62. John Wiley & Sons Ltd.

Van Rijn, L. C. (1993). *Principles of sediment transport in rivers, estuaries and coastal seas.* Aqua Publications, Amsterdam.

Verbeek, H., Kuijper, C., Cornelisse, J. M. and Winterwerp, J. C. (1991). *Deposition of graded natural muds in the Netherlands.* Workshop on Near-shore and Estuarine Cohesive Sediment Transport, Florida, April 1991.

Verreet, G. and Berlamont, J. (1988). Rheology and non-Newtonian behaviour of sea and estuarine mud. *Encyclopedia of Fluid Mechanics, Volume VII, Rheology and non-Newtonian Flows*, ed., N. P. Cheremisinoff, 135–149. Gulf Publishing Company, Houston, Texas.

Wang, Y. W., Larsen, P., Nestmann, F. and Dittrich, A. (1998). Resistance and drag reduction of flows of clay suspensions. *J. Hyd. Engng*, **124**, 41–49. ASCE.

Wellershaus, S. (1981). Turbidity maximum and mud shoaling in the Weser estuary. *Arch. Hydrobiol*, **92**(2), 161–198.

Whitehouse, R. J. S. (1993). Combined flow sand transport: field measurements. *Proc. 23rd ICCE*, **3**, 2542–2555. Venice, ASCE.

Whitehouse, R. J. S. and Roberts, W. (1999). *Predicting the morphological evolution of intertidal mudflats.* HR Wallingford Report SR 538/ Environment Agency R&D Technical Report W191.

Whitehouse, R. J. S., Bassoullet P., Dyer, K. R., Mitchener, H. J. and Roberts, W. (2000). The influence of bedforms on flow and sediment transport over intertidal mudflats. *Continental shelf research*, **20**, 1099–1124. Pergamon Press.

Wilkinson, R. H. (1984). A method for evaluating statistical errors associated with logarithmic velocity profiles. *Geo-Marine Letters*, **3**, 49–52.

Williams, D. J. A. and Williams, P. R. (1989). Rheology of concentrated cohesive suspensions. *J. Coastal Res.*, Special Issue No. 5, 165–174.

Williamson, H. J. and Ockenden, M. C. (1993). Laboratory and field investigations of mud and sand mixtures. In: *Advances in hydro-science and engineering*, Vol. 1, 622–629, ed., S. S. Y. Wang. Washington DC.

Winterwerp, J. C. (1999). On the dynamics of high-concentrated mud suspensions. PhD Thesis published as Report No. 99-3, *Communications on hydraulic and geotechnical engineering*. Delft University of Technology, Faculty of Civil Engineering and Geosciences.

Winterwerp, J. C., Uittenbogaard, R. E. and de Kok, J. M. (2000). Rapid siltation from saturated mud suspensions. In: *Coastal and estuarine fine sediment transport: processes and applications*, eds, W. H. McAnally and A. J. Mehta. Elsevier, Amsterdam. In press.

Wolanski, E., Gibbs, R. J., Mazda, Y., Mehta, A. and King, B. (1992). The role of turbulence in the settling of mud flocs. *J. Coastal Res.*, **8**(1), 35–46.

Young, R. N. and Southard, J. B. (1978). Erosion of fine-grained marine sediments: sea floor and laboratory experiments. *Geol. Soc. Am. Bull*, **89**, 663–672.

# Index

Page numbers in *italic* refer to figures